Lecture Notes in Earth Sciences 64

Editors:
S. Bhattacharji, Brooklyn
G. M. Friedman, Brooklyn and Troy
H. J. Neugebauer, Bonn
A. Seilacher, Tuebingen and Yale

Springer
Berlin
Heidelberg
New York
Barcelona
Budapest
Hong Kong
London
Milan
Santa Clara
Singapore
Paris
Tokyo

Aronne Armanini
Masanori Michiue (Eds.)

Recent Developments on Debris Flows

Springer

Editors

Prof. Dr. Aronne Armanini
Dept. of Civil Environmental Engineering
University of Trent
Via Mesiano di Povo, 77
I-38050 Trent, Italy

Prof. Dr. Masanori Michiue
Dept of Civil Engineering, Tottori University
101 Minami – 4 Koyama
Tottori 680, Japan

Text Editing and Layout

Federica Pedrotti
Department of Civil Environmental Engineering
University of Trent
Via Mesiano di Povo, 77
I-38050 Trent, Italy

"For all Lecture Notes in Earth Sciences published till now please see final pages of the book"

Cataloging-in-Publication data applied for

Die Deutsche Bibliothek - CIP-Einheitsaufnahme

Recent developments on debris flows / Aronne Armanini ; Masanori Michiue (ed.). - Berlin ; Heidelberg ; New York ; Barcelona ; Budapest ; Hong Kong ; London ; Milan ; Paris ; Santa Clara ; Singapore ; Tokyo : Springer, 1997
(Lecture notes in earth sciences ; 64)
ISBN 3-540-62466-X
NE: Armanini, Aronne [Hrsg.]; GT

ISSN 0930-0317
ISBN 3-540-62466-X Springer-Verlag Berlin Heidelberg New York

Typesetting: Camera ready by F. Pedrotti
SPIN: 10492908 32/3142-543210 - Printed on acid-free paper

Preface

This work collects the recent papers discussed in the International Workshop on Debris Flows held in Kagoshima, Japan, in 1993, in response to the International Decade of Natural Disaster Reduction Program. Although the debris flow was called as a shadowy disaster about thirty years ago which often killed a few hundreds people, the recent studies are going to clarify the mechanism of the flow and the occurrence, and moreover the behaviour of debris flow disaster.

In order to mitigate and prevent debris flow disasters, we should forecast the location and time for the occurrence of debris flow due to rainfall through the theoretical and empirical approaches. These forecasting methods can contribute to make plans of evacuation and to evacuate in real time inhabitants in mountain areas.

Debris flow may be defined as a hyper-concentrated flow of the mixture of water and sediment. Therefore, the characteristics of the flow are determined by the concentration, the grain size and hydraulic conditions such as velocity and depth. Readers will recognize that the choice of dominant facts is different by researchers when modelling the shear stress of debris flow.

Check dams have been developed as main control measures for the debris flow. The design of these structures relied mostly on field observation and experience. However, a check dam of new type such as a slit dam which catches boulders transported by debris flow but flows out the sediment of small grain size due to bed load transportation has been recently developed through the observation and the dynamic of debris flow. Some papers report examples of these countermeasures of new concepts against debris flow, the design of effective check dams, and the field survey for debris flow.

Finally, I hope that you will be interested in the phenomenon of debris flow through the book, and also the disaster of debris flow will be mitigated by the development of research on debris flow.

Masanori Michiue

Acknowledgements

The Editors of the book are indebted to the Scientific Committee of the Workshop:

Prof. H.W. Shen (University of California, U.S.A.)
Prof. G. Di Silvio (University of Padova, Italy)
Dr. L. Ding (IRTICES, China)

and to the Local Organizing Committee:

Prof. M. Hirano (Kyushu University)
Prof. T. Takahashi (Kyoto University)
Prof. E. Shimokawa (Kagoshima University)
Prof. T. Mizuyama (Kyoto University)
Dr. K. Miyamoto (Sabo Technical Center)

Table of Contents

Chapter 3: Control Measures for Debris Flow

List of Contributors

Aguirre-Pe Julian — *Laboratorio de Hidráulica, Facultad de Ingegneria, Universidad de Los Andes, Apartado 45, Mérida 5101-A, Venezuela*

Armanini Aronne — *Department of Civil and Environmental Engineering, University of Trent, Italy*

Davies Tim R. — *Department of Natural Resources Engineering, Lincoln University, New Zealand*

Egashira Shinji — *Disaster Prevention Research Institute, Kyoto University, Gokasho, Uji, Kyoto 611, Japan*

Hashimoto Haruyuki — *Department of Civil Engineering, Kyushu University, Fukuoka 812, Japan*

Hirano Muneo — *Department of Civil Engineering, Kyushu University, Fukuoka 812, Japan*

Ikeya Hiroshi — *Sediment Control Department, Ministry of Construction, 2-1-3, Kasumigaseki, Chiyoda-ku Tokyo, Japan*

Ishikawa Yoshiharu — *Erosion Control Department, Public Works Research Institute, Ministry of Construction, 1-Asahi, Tsukuba 305, Japan*

Jaeggi Martin N.R. — *Laboratory of Hydraulics, Hydrology and Glaciology, Federal Institute of Technology, Zurich, Switzerland*

Jan Chyan-Deng — *Department of Hydraulics and Ocean Engineering, National Cheng Kung University Tainan, Taiwan 70101, R.O.C.*

Jitousono Takashi — *Faculty of Agriculture, Department of Environmental Sciences and Technology, Kagoshima University, Kagoshima 890, Japan*

Julien Pierre Y. — *Engineering Research Center, Colorado State University, Fort Collins, CO 80523, USA*

Kitamura Ryosuke — *Faculty of Engineering, Department of Ocean Civil Engineering, Kagoshima University, Kagoshima 890, Japan*

Michiue Masanori *Department of Civil Engineering, Tottori University, Tottori, Japan*

Mizuyama Takahisa *Laboratory of Erosion Control, Faculty of Agriculture, Kyoto University, Kyoto 606-01, Japan*

Shen Hsieh Wen *Department of Civil Engineering , University of California, Berkeley, CA 94720, USA*

Takahashi Tamotsu *Disaster Prevention Research Institute, Kyoto University, Gokasho, Uji, Kyoto 611, Japan*

O'Brien Jim S. *Hydraulic Engineer, FLO Engineering, Inc., P.O. Box 1659, Breckenridge, CO 80424, USA*

Okubo Shun *Sediment Control Department, Ministry of Construction, 2-1-3, Kasumigaseki, Chiyoda-ku Tokyo, Japan*

Pellandini Stefano *Laboratory of Hydraulics, Hydrology and Glaciology, Federal Institute of Technology, Zurich, Switzerland*

Shimokawa Etsuro *Faculty of Agriculture, Department of Environmental Sciences and Technology, Kagoshima University, Kagoshima 890, Japan*

Taniguchi Yoshinobu *Department of Agriculture and Forest Sciences, Miyazaki University 889-21, Japan*

Yamada Takashi *Erosion Control Department, Public Works Research Institute, Ministry of Construction, 1-Asahi, Tsukuba 305, Japan*

Chapter 1

Observation and Measurement for Debris Flow

Introduction

Etsuro Shimokawa

1. Introduction

A debris flow consists of three processes of initiation, flow and deposition. Many field observations and measurements have been carried out to examine factors controlling each process, to clarify mechanism and geomorphological effects of each process and as results of those to predict when, where and how magnitude debris flows may occur in many countries in the world. Most the field observations and measurements for debris flow, however, don't considerably contribute to quantitative explanation of the dynamic behaviour of debris flow. Because most the data from the field observations and measurements are static records of debris flow scar which was formed at a long and narrow area from the source through flow and to deposition after the completion.

The field observations and measurements for debris flow made a rapid progress during the last 20 years, which was mainly supported by direct observations and measurements of dynamic motion of debris flow using electronic instruments, such as video camera, ultrasonic-waves water-gauge and ultrasonic-waves current-meter. In this paper, the field observations and measurements for debris flow are briefly outlined in three processes.

2. Initiation process

Initiation of debris flows has some variations with landsliding on a hillslope, scouring of debris mass on a steep valley bed, rilling and gullying on a bare terrain and their composite in origin, depending on geomorphological and hydrological characteristics of the source, geotechnical properties of debris mass mobilized into a debris flow from the source and rainfall or rarely earthquake as a trigger. Most the debris flows would be initiated by sliding or slumping of debris mass on hillslopes and steep valley beds, which are caused by saturation with water. In this type, a source scar of debris flow may be clearly formed. Such scar can be used to assume the initiation mechanism and at the same time to observe and measure geometry, volume and catchment area of the source, and profile face, thickness and components of debris mass at the source as geomorphological and hydrological characteristics. Also texture, density, infiltration capacity and shearing strength of the debris mass as the geotechnical and hydrological properties have been measured by using the debris mass samples collected from the scar and its surroundings (Johnson 1984).

It is important for predicting debris flow disasters to know where and how volume are the potential debris mass on hillslopes and/or valley beds, which is mobilized into debris flow from the source. However, a few geomorphological surveys on them have been done at some areas in which debris flow disasters have often occurred (Suwa and Okuda 1988). Rainfall observation is also indispensable to examine the initiation of debris flow and to predict when debris flow may occur as it is mostly triggered by a storm.

In volcanic areas, debris flows or volcanic mud flows have often occurred. The volcanic debris flows are laughly divided into two types in initiation, primary one associated with eruptions and secondary one indirectly associated with eruptions. The primary one is caused by a eruption through a crater lake, melting of snow and ice due to eruption products and descent of pyroclastic flows into streams. On the other side, the secondary one is caused by the following four, an increase of surface runoff by tephra covering, a rapid melting of snow and ice, collapse of crater lake impoundment and earthquakes (Blong 1984). Of these, the debris flows caused by the increase of surface runoff have been often surveyed in many volcanoes, Irazu (Waldron 1967), Usu (Kadomura et al. 1983), Mt.St.Helens (Pierson 1986), Sakurajima (Shimokawa and Jitousono 1994 in this volume).

3. Flow process

Debris flow may erode valley bottom and side and transport much sediment during the motion. A scar like a snake is formed on the channel whose debris flow passed through. Superelevation of the flow appears along a transverse section around a bend of the channel by the centrifugal force. These scar are effective to measure geometrical characteristics and deformation of channel by debris flow and to estimate hydrological and hydraulic properties of a debris flow, such as velocity, a peak stage, a peak discharge and impact pressure.

Direct observations and measurements of the flow motion in fields firstly started at Yakedake in Japan by Okuda et al. (1980) of Kyoto University cooperated the Japanese Ministry of Construction in the early 1970s, being followed at Sakurajima in Japan (Watanabe and Ikeya 1981), Dongchuan in China (Kang, Z. and Tang, B. 1985) and Mt. St. Helens in USA (Pierson 1986). Main equipments in the observation system of debris flow by Okuda et al. (1980) consist of constant interval shot camera and video camera for taking static and dynamic pictures of a debris flow and wire sensors for detecting occurrence of a debris flow. The observation system is automatically operated when the sensors detect the occurrence and a front of debris flow came within a sight of the cameras, as it is impossible to foresee when debris flow may occur. Recently, ultrasonic-waves water-gauge and ultrasonic-waves current-meter as high-technical instrument in addition to their instruments have been used to measure the hydrological and hydraulic properties (Shimokawa and jitousono 1994 in this volume). These instruments are effective for observing the hydrological and hydraulic properties of debris flow throughout the year without detecting of debris flow occurrence by the sensors. Sometimes, dip samples of debris flow slurry during the motion were collected to measure sediment concentration and particle size distribution at a few observation sites (Watanabe and Ikeya 1981, Pierson 1986). Many data obtained by these direct observations and measurements created a significant contribution to understanding and theoretically analyzing the mechanical behaviour of the flow motion in fields.

4. Deposition process

Although the debris flow may leave its transported sediment around a bend and/or gentle and wide part of channel even during the motion, most the sediment is transported to a fan. Coming down to the fan, a debris flow would decelerate to lastly cease and consequently leave deposits of sediment including large boulders on the fan surface. The deposits are divided into two, lateral deposits (or levee) which are formed along one or both sides of the flow during the motion and terminal ones which are formed when the debris flow stop to move (Suwa 1989). The typical terminal deposits would form tonge-shaped fronts or lobes with large boulders. Field observations and measurements are aimed to clarify the mechanism and geomorphological effects of depositional process of debris flow and to predict of potential hazard area caused by debris flow. The field observations and measurements are mostly concerned with geomorphological and geotechnical characteristics of the deposits and geomorphological effects of debris flow on the fan formation. A dynamic direct observation on the depositional process were carried out at Yakedake, using the video camera and other equipments which are automatically operated as mentioned above. This observation enabled us a analysis of the mechanical behaviour on the motion of flow front and large boulders in the depositional process (Okuda, S. et al. 1981).

REFERENCES

Blong, R. J. (1984): Volcanic hazards, 424pp., Academic Press, Sydney.

Johnson, A. M.(1984): Debris flow. In Brunsden D. and Prior D. P. ed., Slope Instability, Chichester, Johen Wiley & Sons, pp. 257-361.

Kadomura, H., Imagawa, T. and Yamamoto, H. (1983): Eruption-induced rapid erosion and mass movements on Usu volcano, Hokkaido. Zeit. Geomorph. N. F., Suppl. Bd.46, 123-142.

Kang, Z. and Tang, B. (1985): The debris flow and its observation systems in the Jiangjia Gully, Dongchuan, Yunnan. Proceedings. 4th International Conference and Field Workshop on Landslides, Tokyo, pp. 385-390.

Okuda, S., Suwa, H., Okunishi, K. Yokoyama, K. and Nakao, M. (1980): Observation on the motion of a debris flow and its geomorphological effects: Zeit. Geomorph. N. F., Suppl. Bd.35, 142-163.

Okuda, S., Suwa, H. Okunishi, K. and Yokoyama, K. (1981): Depositional processes of debris flow at Kamikamihori Fan, Northern Japan Alps. Transactions Japanese Geomorphological Union 2-2, 353-361.

Pierson, T. C. (1986): Flow behavior of channelized debris flows, Mount St. Helens, Washington. In Abrahams, A. D. ed., Hillslope Processes, Boston, Allen & Unwin, 269-296.

Shimokawa, E. and Jitousono, T. (1994): Field survey for debris flow in volcanic area. Proceedings International Workshop on Debris Flow, IAHR, 101-110.

Suwa, H. and Okuda, S (1988): Seasonal variation of erosional processes in the Kamikamihori valley of Mt. Yakedake, Northern Japan Alps. Catena Suppl. 13, 61-77.

Suwa, H. (1989): Field Observation on Debris flow. Proceedings Japan-China (Taipei) Joint Seminar on Natural Hazard Mitigation, Kyoto, 343-352.

Waldron, H. H. (1967): Debris flow and erosion control problems caused by the ash eruptions of Irazu Volcano, Costa Rica. U.S. Geological Survey Bull. 1241-I, 1-37.

Watanabe, M. and Ikeya, H. (1981): Investigation systems and analysis on volcanic mud flow in Mt. Sakurajima, Japan, in Erosion and Sediment Transport Measurement. International Association of Hydrological Sciences Publication 133, pp.245-256.

Prediction of Debris Flow for Warning and Evacuation

Muneo Hirano

Department of Civil Engineering
Kyushu University
Fukuoka 812, Japan

Abstract

The occurrence condition of debris flow due to rainfall is given by solving the equations for flow on a slope. The solution shows that a debris flow will occur on a slope when the accumulated rainfall within the time of concentration exceeds a certain value determined by the properties of the slope. To estimate this critical value, the system analysis technique would be commendable. In this study, a procedure to find the critical rainfall from the rainfall data with and without debris flows is proposed. Reliability of this method is verified by applying to the debris flows in Unzen Volcano which recently began to erupt.

Discharge of debris flow in a stream is obtained by solving the equation of continuity using the kinematic wave theory and assuming the cross sectional area to be a function of discharge. The computed hydrographs agree well with the ones observed at the rivers in Sakurajima and Unzen Volcanoes. It is found from the derived equation that the runoff intensity of debris flow is in proportion to the rainfall intensity and accumulated rainfall, jointly. This gives a theoretical basis to the conventional method which has been widely used.

1. Introduction

The debris flow has been feared for its potential to cause heavy disaster. Studies on occurrence and intensity of debris flow, therefore, required to prevent the disasters. In the past, occurrence criteria of debris flow have been defined by two parameters, cumulative rainfall from its beginning and a rainfall just before the occurrence of debris flow. But this method is not satisfactory in accuracy as well as in deciding the cumulative rainfall in practice because of the lack of theoretical clarity.

In this paper, the occurrence conditions of debris flow are analyzed to obtain the critical rainfall needed to cause a debris flow, and a mathematical model of debris flow runoff which predicts the intensity of debris flow is derived.

2. The Critical Rainfall for Occurrence of Debris Flow

2.1 Occurrence Criteria of Debris Flow

On a slope of deposits shown in Fig. 1, the shear stress τ at a point in the deposit is given by

$$\tau = \{C_*(\sigma - \rho)a + \rho(h_0 + a)\}g \sin\theta \tag{1}$$

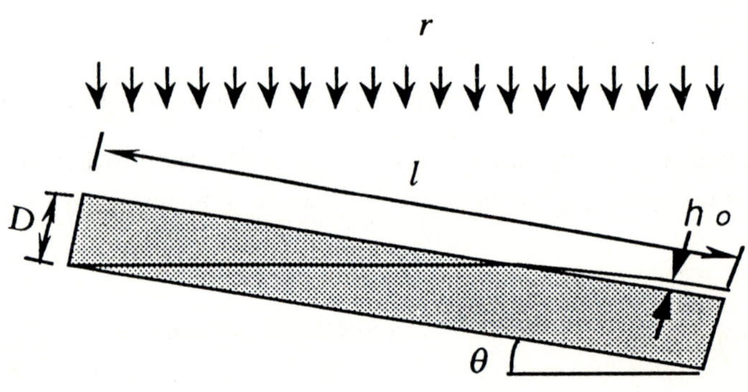

Fig. 1. Schematic sketch of a slope

where, C^* is the concentration of deposited material, σ and ρ are the density of the deposits and water, respectively, a is the distance from the surface, h_o is the depth of the surface flow, g is the gravitational acceleration and θ is the angle of the slope. The resisting stress τ_L at the point is expressed as

$$\tau_L = c + C_*(\sigma - \rho)ag\cos\theta\tan\phi \tag{2}$$

where c is the adhesive force, and ϕ is the angle of internal friction.

Since the critical condition is $\tau = \tau_L$, the critical angle of a slope θ_c for the occurrence of a debris flow is obtained by Eqs. (1) and (2) as

$$\tan\theta_c = \frac{c/(\rho ga\cos\theta_c) + c_*(\sigma/\rho - 1)\tan\phi}{c_*(\sigma/\rho - 1) + 1 + h_0/a} \tag{3}$$

By substituting ordinary values of $C_* = 0.6$, $tan\phi = 1.0$, $\sigma/\rho = 2.65$ and $c = 0$ for sandy materials to Eq. (3) and considering that a and h_o should be larger than grain size d to cause a debris flow [1], one obtains $\theta_c = 14.8^o$. This has been supported by field data as well as flume data.

2.2 Critical Rainfall for Occurrence of Debris Flow

According to the theory mentioned above, a debris flow will occur on a slope deeper than θ_c qc when depth of the surface flow exceeds the grain size. There are two approaches to obtain the critical rainfall based on this theory.

One is to give the discharge of surface flow in which the depth is equal to the grain diameter of the deposits as the critical discharge. Ashida et al. [2] derived the critical discharge Q_c by putting $h_o = d$ and $Q_c = Buh_o$ as

$$Q_c = \sqrt{\frac{8\sin\theta}{f_o\kappa^3}B^2 g\, d^3} \tag{4}$$

where, B is the width of the flow, u_c is the velocity of surface flow, f_o is the resistance coefficient, κ is the ratio of h_o and d close to unity and d is the grain diameter of the deposits.

Applying the rational formula to Eq. (4), one obtain the critical rainfall intensity as

$$r_T = \frac{1}{T}\int_0^T r\,dt \geq \frac{B\,d}{f\,A}\sqrt{\frac{\sin\theta}{f_o\kappa^3}\,g\,d} \tag{5}$$

where, T is the time of concentration, f is the runoff coefficient, and A is the catchment area.

The other is to assume the occurrence of surface flow to be the occurrence condition of debris flow. Since irregularity of the slope surface is larger than the grain size, depth of the surface flow will exceed in some part of the slope when surface flow appears on the slope. Consequently, a debris flow will occur as soon as surface flow appears on a slope due to the heavy rainfall. The criteria for the surface flow are given as follows:

On a slope shown in Fig. 1, the momentum and continuity equations of subsurface flow are expressed by

$$\frac{\partial\lambda h}{\partial t} + \frac{\partial v h}{\partial x} = r\ \cos\theta \qquad \text{and} \qquad \mathrm{v} = \mathrm{k}\ \sin\theta \tag{6}$$

where, λ is the porosity, h is the depth of the subsurface flow, t is the time, v is the velocity of the flow, x is the coordinate taken in the downstream direction, r is the rainfall intensity, and κ is the hydraulic conductivity.

By solving Eq. (6) by using the kinematic wave theory, one obtains the occurrence conditions of surface flow as

$$l \geq kT\ \sin\theta/\lambda \qquad \text{and} \qquad \lambda\mathrm{D} \geq \int_0^{\mathrm{T}} \mathrm{r}\ \cos\theta\ \mathrm{dt} \tag{7}$$

where, l is the length of the slope, T' is the time of concentration, and D is the depth of the deposits.

Assuming that debris flow occurs when surface flow appears on a slope, the occurrence condition of debris flow is derived from Eq. (7) as

$$r_T = \frac{1}{T}\int_0^T r\, dt \geq \frac{D\,k}{l}\tan\theta \tag{8}$$

The applicability of this equation was verified by the experiment [3] as shown in Fig. 2.

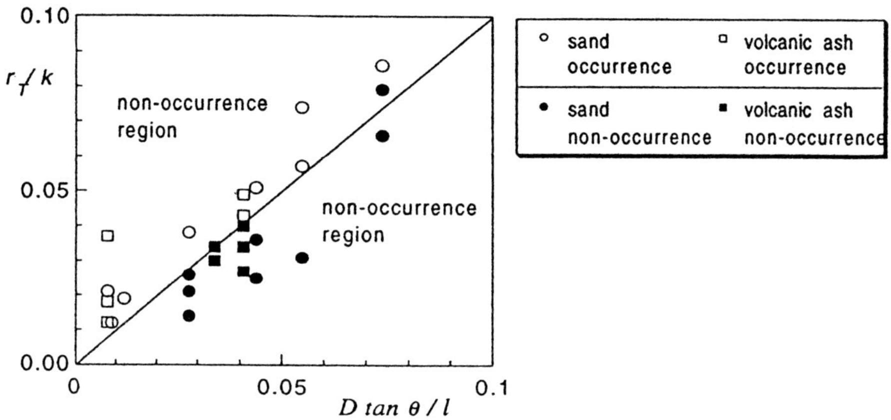

Fig. 2. Comparison between theoretical and experimental results

In spite that Eqs. (5) and (8) are derived from the different basis, right hand sides of the equations are the same. These equations indicate that a debris flow will occur when rainfall intensity within the time of concentration exceeds a certain value determined by the properties of the slope.

2.3 Estimation of the Critical Rainfall

2.3.1 Estimating Method. Equation (8) is rewritten as

$$R(t,T) = \int_{t-T}^{t} r(\tau)\, d\tau \geq \frac{D\,k}{l}\tan\theta = R_c \tag{9}$$

where, t is the time, and R_c is the critical rainfall. Equation (9) shows that debris flow will occur when cumulative rainfall within the time of concentration exceeds a certain value related to the properties of the slope. Two parameters, the time of concentration T and critical amount of rainfall R_c , should be estimated to obtain the criterion for occurrence of debris flow. It may be possible to estimate the value of R_c by measuring the value of D, l and q, however, the estimated value will not be accurate enough for practical use due to the large errors in the measurements. This is the reason why the method of system analysis will be commendable to identify the parameters.

To estimate the time of concentration and critical rainfall, T and R_c , cumulative rainfall $R(t, t_o)$ defined as below is calculated.

$$R(t, t_o) = \int_{t-t_o}^{t} r(\tau)\, d\tau \tag{10}$$

The maximum values of $R(t, t_o)$ for each time, $R_{max}(t_o)$, are plotted against t_o. If there are no errors in the data as well as in the theory, the plotted lines should exceed the point $R_{max}(T)$ when debris flow occurred, and not exceed the point when debris flow did not occur as schematically illustrated in Fig. 3. Consequently, the upper limit line of non-occurrence and the lower limit line of occurrence should cross at the point $R_{max}(T)$ as schematically shown in Fig. 4(a). Because of the errors in the data and the unsteady field conditions, however, the upper limit of non-occurrence and the lower limit of occurrence will be like two lines shown in Fig. 4(b). The point where the difference between two curves is minimum is estimated to be the time of concentration.

2.3.2 The critical rainfall of debris flow in the Mizunashi River. Unzen volcano began to erupt in November 1990 after 198 years of dormancy and has been in violent activity. Continuous growth of lava dome and falls of

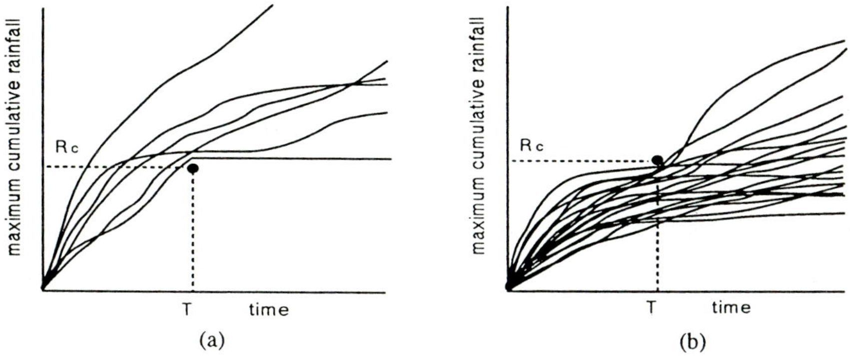

Fig. 3. Cumulative rainfall when debris flow occurred(a) and not occurred(b)

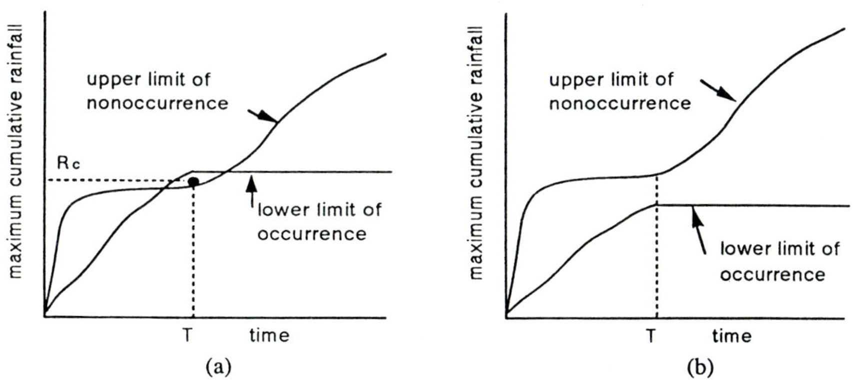

Fig. 4. Upper limit of non-occurrence and lower limit of occurrence

lava rocks have resulted in frequent pyroclastic flows. As a great volume of volcanic material has been deposited and scattered by the pyroclastic flows, debris flows have frequently occurred along the Mizunashi River and damaged many houses.

The cumulative amounts of rainfall were calculated using the rainfall data collected by the Unzen Meteorological Observatory, both when debris flows had occurred and when they had not. In cases when debris flows occurred, the amount of rainfall up until the time of occurrence was computed, and in cases without debris flows, whole data were used. In Fig. 4, the upper limit of non-occurrence and lower limit of occurrence are illustrated.

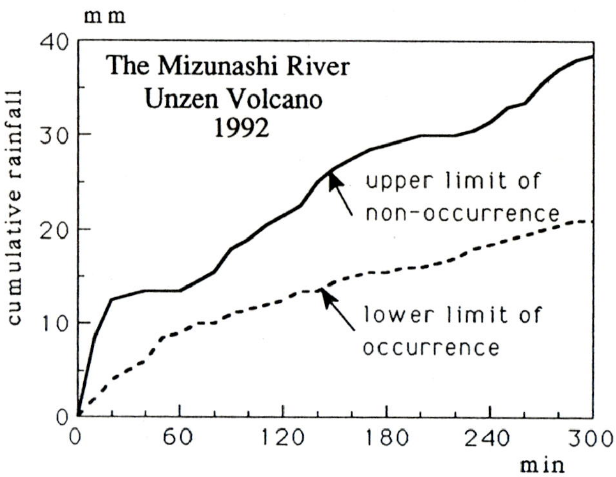

Fig. 5. Upper limit of non-occurrence and lower limit of occurrence of debris flow in the Mizunashi River, Unzen Volcano

From the Fig. 4, the following are confirmed: 1) the time of concentration is estimated to be about an hour on average; 2) the occurrence of debris flows is possible when the amount of rainfall per hour rises over the limit of 9 mm; and 3) debris flows will definitely occur when this amount rise over the limit of 14 mm. At Volcano Sakurajima, which has been in violent activity in this

20 years, debris flows have been generated by rainfall from 7 to 13 mm over a period of forty minutes. By comparison, the debris flows in the Mizunashi River show the typical property of volcanic debris flow which is possible by a small amount of rainfall.

3. Runoff Analysis of Debris Flow

3.1 Runoff Coefficient of Debris Flow

Runoff coefficient of the debris flow, f, is defined as the ratio of the flow rate and the rainfall intensity as

$$f = \frac{flow\ rate}{(rainfall\ intensity) \times (catchment\ area)} = f_s\, F \tag{11}$$

where

$$f_s = \frac{flow\ rate}{(rainfall\ intensity) \times (area\ where\ debris\ flow\ has\ occurred)}$$

and

$$F = \frac{(area\ where\ debris\ flow\ has\ occurred)}{(catchment\ area)}$$

The continuity condition leads the following equation for f_s as [4]

$$f_s = (1 - \lambda)/(1 - \lambda - C) \tag{12}$$

where, C is the concentration of debris flow, and λ is the porosity of the deposits. It is seen in the above equation that the range of fs is unity to infinitive., and for water flow, fs is unity as $C = 0$. According to the experiments [3], $l = 0.54$, C=0.50 and fs =18. In usual runoff of water flow, F is considered to be unity, while in the case of debris flow, F should be less than unity varying with time.

(13)

3.2 Modeling of Runoff

The equation of continuity in a stream is given by

$$\frac{\partial A_s}{\partial t} + \frac{\partial Q}{\partial x} = q_s + q_* \tag{14}$$

where, A_s is the cross sectional area of the stream, Q is the discharge of the flow, and q_s is the lateral inflow rate and q_* is the rate of erosion of bed and bank. Lateral inflow rate is expressed by

$$q_s = f_s r\, l \,\cos\theta \tag{15}$$

Assuming A_s to be a function of Q, Eq. (14) is solved by use of the characteristic curve as follows:

On the characteristic curve $d_x/d_t = dQ/dA_s$,

$$Q = \int_0^L (q_s + q_*) dx \tag{16}$$

Substituting Eq. (15) into Eq. (16) and neglecting the erosion rate, one obtains

$$Q = \int_0^L f_s r\, l \,\cos\theta \, dx \tag{17}$$

If we substitute a constant rainfall intensity r_o into Eq. (17), then we obtain

$$Q = f_s r_o \int_0^L r\, l \,\cos\theta \, dx \tag{18}$$

Considering $Q = f_s r_o A$ in this case, the following is obtained.

$$A = \int_0^L l \cos\theta \, dx \qquad \text{or} \qquad \int_0^L l \cos\theta \, dx/A = 1 \tag{19}$$

As $lcos\theta dx$ is a very small area of a watershed, $lcos\theta dx/A$ is taken to be a probability function of a slope, resulting in the following expression.

$$Q = A \int_0^\infty \int_0^\infty f_s r \, \phi(\eta, l) \, d\eta \, dl \tag{20}$$

where,$\phi(\eta, l)$ is the probability function of $\eta = \lambda D$ and l.

On a slope where the conditions given by Eq. (4) are satisfied, a debris flow will occur. While on a slope where rainfall intensity is less than the critical values given by Eq. (5), water flows into a stream, but no debris flow occurs on the slope.

3.2.1 Runoff Model for Water Flow. When all slopes are shorter and/or thicker than the critical values given by Eq. (4), no debris flow will occur in the watershed. In this case, $f_s = 1$ and rainfall intensity is defined as

$$r_T = \frac{1}{T} \int_{t-T}^{T} r(\tau) \, d\tau \tag{21}$$

Substitution of Eq. (21) into Eq. (19) yields

$$Q(t) = A \int_0^\infty \int_0^\infty r_T \, f(l) \, dl = A \int_0^\infty r_T \, \Phi(T) \, dT \tag{22}$$

where,

$$f(l) = \int_0^\infty \phi(\eta, l) \, d\eta \qquad \text{and} \qquad \Phi(T) = f(l) \frac{dl}{dT}$$

From Eq. (22), instantaneous unit hydrograph $u(t)$ is derived as

$$Q = A \int_0^\infty r(t-\tau)\, u(\tau)\, d\tau \tag{23}$$

where,

$$u(\tau) = \int_0^\infty \frac{1}{T} \Phi(T)\, dT \tag{24}$$

It is clarified that instantaneous unit hydrograph is a function of the time of concentration.

3.2.2 Runoff Model for Debris Flow. From Eq. (4), debris flow occurs on a slope where the followings are met.

$$l \geq \frac{k\ \sin\theta}{\lambda}(t - t_o) \qquad \text{and} \qquad \lambda D = \int_{t_o}^{t} r\ \cos\theta\ dt \tag{25}$$

Applying Eq. (25) to Eq. (20), the discharge of debris flow is expressed by

$$\begin{aligned} Q_s(t) &= \sum_{t_o=-\infty}^{t} f_s\, r\, \phi(\eta, l)\, \Delta\eta\ \Delta l \\ &= f_s\, r \left\{ \int_{kt \sin\theta/\lambda}^{\infty} \phi(\eta_o, l)\ dl\ \Delta\eta_o + \sum_{t_o=0}^{t} \phi(\eta, l)\, \Delta\eta\ \Delta l \right\} \end{aligned} \tag{26}$$

where:

$$\eta_o = \int_0^t r\ \cos\theta\ dt \quad , \quad \eta = \int_{t_o}^{T} r\ \cos\theta\ dt \quad \text{and} \quad l = k\, t\ \sin\theta/\lambda \tag{27}$$

From the relations mentioned above, one obtains

$$\Delta\eta_o = \Delta\eta = r\ \cos\theta\, \Delta t \qquad \text{and} \qquad \Delta l = k\, t\, \Delta\ \sin\theta/\lambda \tag{28}$$

Assuming that debris on a slope outflows in a short period of time Δt, one obtains

$$\int_{t}^{t+\Delta t} q_s\,dt = \int_{t}^{t+\Delta t} f_s r\,l\,\cos\theta\,dt = Dl + \int_{t}^{t+\Delta t} r\,l\,\cos\theta\,dt. \tag{29}$$

and

$$\Delta t = \frac{D}{(f_s-1)\,r\,\cos\theta} \tag{30}$$

Flow rate of debris flow is obtained by substituting Eqs.(27)-(30) into Eq. (26) as

$$\begin{aligned} Q(t) &= A\,r(t)\frac{f_s}{(f_s-1)\lambda}\left\{\eta_o \int_{kt\sin\theta/\lambda}^{\infty} \phi(\eta_o, l)\,dl \right. \\ &= \left. \frac{k\,\sin\theta}{\lambda}\int_{0}^{t} \eta\phi\left(\eta, \frac{k(t-t_0)\sin\theta}{\lambda}\right) dt_0 \right\} \end{aligned} \tag{31}$$

Assuming that λD and l are independent each other,

$$\phi(\eta, l)\,d\eta\,dl = \phi_\eta\,\phi_l\,d\eta\,dl \tag{32}$$

where, ϕ_η and ϕ_l are the probability function of λD and l, respectively.

Since the first term of the right hand side of Eq. (31) is dominant compared with the second one, Eq. (31) can be simplified considering Eq. (32) as

$$Q(t) = A\,r(t)\frac{f_s}{(f_s-1)\lambda}\eta_o\phi_\eta(\eta_o)\int_{kt\sin\theta/\lambda}^{\infty} \phi_l(l)\,dl \tag{33}$$

It has been found that distribution of slope length in a watershed is likely to be log-normal [5]. Results of application of Eq. (33) to debris flows in Sakurajima and Unzen Volcanoes are shown in Fig. 6.

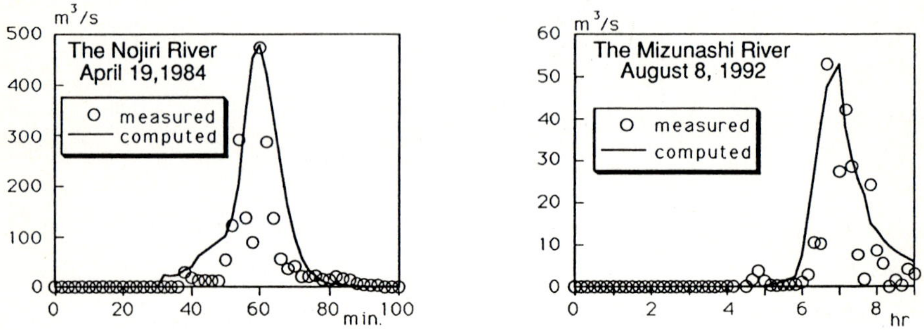

Fig. 6. Comparison between computed and observed hydrographs

3.3 Prediction of Debris Flow

Equation (33) is rewritten as

$$\frac{Q(t)}{A} = r(t)\,\eta_o M \tag{34}$$

where,

$$M = \frac{f_s}{(f_s - 1)\lambda}\phi_\eta(\eta_o) \int_{kt\sin\theta/\lambda}^{\infty} \phi_l(l)\,dl \qquad \text{and} \qquad \eta_o = \int_0^t r\cos\theta\,dt$$

Equation (34) indicates that the runoff intensity of debris flow is in proportion to the rainfall intensity at the time $r(t)$ and the cumulative rainfall up to that time η_o, jointly. This means that a constant value of $Q_s/(AM)$ is shown as a hyperbola on a $[r(t), \eta_o]$ plane as schematically illustrated in Fig. 7.

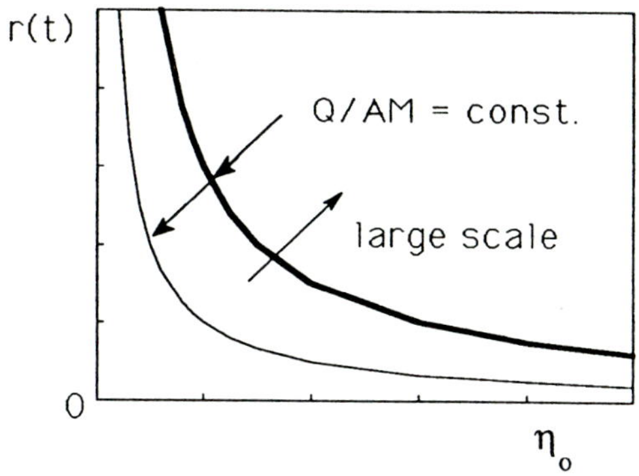

Fig. 7. Schematic sketch of diagrams for forecasting debris flow

An empirical method by use of hyperbola-like curve(s) on the $[r(t), \eta_o]$ plane has been widely used to forecast the occurrence of debris flow. Equation (33) indicates that this conventional method has a theoretical basis and to be useful to predict the intensity of debris flow but not predict occurrence itself.

4. Conclusions

The occurrence condition of debris flow due to heavy rainfall and runoff analysis of debris flow were studied. Results obtained are as follows:

1. Debris flow will occur on a slope when amount of rainfall within the time of concentration exceeds a certain value which is peculiar to the slope. The time of concentration and the critical amount of rainfall is obtainable by analyzing the data of rainfall and debris flows.
2. A mathematical model for runoff of debris and water flows is derived. This model results in the instantaneous unit hydrograph when no debris

flow occurs. The applicability of this model was verified by adopting to the debris flows in Sakurajima and Unzen Volcanoes.

3. The derived equation for debris flow discharge gives the theoretical basis to an empirical method which uses the cumulative rainfall and rainfall intensity at the moment to forecast the occurrence of debris flow.

REFERENCES

[1] Takahashi, T.(1977): A mechanism of occurrence of mud-debris flow and their characteristics in motion, Disaster Prevention Institute Annuals, No.21 B-2, pp.405-435 (in Japanese).

[2] Ashida, K., T. Takahashi and K. Sawai(1978): Evaluation of risk due to debris flows, Disaster Prevention Institute Annuals, No.22 B-2, pp.423-439 (in Japanese).

[3] Hirano, M., M. Iwamoto and T. Harada(1976): Study on the mechanism of occurrence of debris flow by artificial rainfall. Preprints of the Annual meeting of JSCE. pp.299-301, (in Japanese).

[4] Hirano, M., T. Moriyama, M. Hikida and M. Iwamoto(1985): A modeling of debris flow in the active volcanic area. Proceedings of International Symposium on Erosion, Debris Flow and Disaster Prevention, Tsukuba, pp.265-270.

[5] Hirano, M.(1983): Modeling of runoff process in a first-order basin. Journal of Hydroscience and Hydraulic Engineering, Vol. 1, No. 2, pp.113-123.

DISCUSSION

Aguirre Pe: Resistance coefficient f_o in Eq. (4) is related to friction between clear water and the saturated material?

Hirano: Yes, f_o is the friction coefficient of a clear water flow on a bed. In Eq. (4), Q_c is the discharge of a clear water flow just before a debris flow occurs on the slope.

Mizuyama: The system analysis assumes that conditions of basins or torrents do not change much. In active volcanoes, the depth of pyroclastic deposit, gully networks and infiltration rate have changed much. I think that prediction methods should include such things as parameters.

Hirano: There are many system analysis methods which are applied to unsteady systems. Classic methods such as Wiener-Hopf equation deal only with the steady system, but recent methods such as Kalaman filtering (1960), neural networks, etc. have been developed to apply the unsteady systems. This is one of the reasons we introduced neural networks. The advantage of neural networks is that any kind of data can be used as input data. If the data of infiltration rate, hydraulic conductivity, etc. are available, these data can easily be used in the prediction by using the neural networks. But how can we obtain those data at the mountain side of Unzen Volcano where pyroclastic flows have frequently occurred? System analysis is a different approach from the deterministic ones. All factors should not always be include in the analysis, as is seen in most of all runoff analysis. It is also noticed that there is no significant change in the occurrence criteria of debris flows at the Mizunashi River, in spite of the fact that gully networks and other geological and topographical characteristics are much changed.

Davies: 1. Debris flow occurs and travels very quickly in small catchments. Therefore, warning time is very short (about 1/2 hour or 1 hour). Is this long enough to evacuate people - especially old people, at night, in bad weather?

2. Fig. 6 shows several measured surges in each river but Eq. (33) predicts only one hydrograph peak in each river. Perhaps your model only works for single debris flow surge? Also, if several surges did occur, computed rinoff volume is much too small?

Hirano: 1. One hour may not be long enough, therefore, the prediction of rainfall should be required for the effective evacuation. Weather forecasting made by Meteorological Observation should be considered in warning and evacuation.

2. We applied this model to a debris flow which had binomial peaks of discharge, and confirmed that the computed hydrograph has also binomial peaks. This method will be applicable to the flows with several peaks, when these peaks were caused by the several peaks in rainfall. But the several surges in Fig. 6 might be caused by the large fluctuations of the data of velocity and water levels. This model will not be able to follow these fluctuations on hydrograph.

Takahashi: The time of concentration you referred seems to me too long beacuase the catchment area of the debris flow occurrence is very small.

Hirano: Two kinds of time of concentrations have been used in runoff analysis, one is the time of concentration on a slope and the other is that in a stream. In a small watershed, the former one is generally much longer than the latter one. I think you meant the time of concentration in a stream. In the Mizunashi River, the time of concentration of the stream is estimated to be 10 to 20 minutes from the records of the seismograph and wire-sensors. The value we obtained is the time of concentration on the slope, but not of the stream.

Egashira: Is it important to specify the concentration time of rainfall to predict the occurrence of debris flow in terms of neural network method? Because the judgement of the occurrence of debris flow may depend on place to place of the river reach.

Hirano: In some cases, it is difficult to estimate the time of concentration. It is the main reason why we introduced the neural networks. In this study, the time of concentration is not used as the input to the neural networks. This model is also useful to estimate the time of concentration.

Julien: Intuitively, debris flow should occur under heavy rainfall r_t and steep slopes (large θ). Can you please clarify Fig. 2 given the threshold of occurrence shown by the line at 45°. For a given κ, D and l constant, the domain of occurrence for heavy rainfall is the upper left triangle. For a given κ, D and l constant, the domain of occurrence for steep slopes in the lower right triangle. Those views are incompatible!

Hirano: If one substitutes h_o=0, C^*=0.6, tanϕ=0.8, σ/ρ=2.65 and c=0 to Eq. (3), one obtains θ=21.7^o. Therefore, on a slope steeper than 22°, a debris flow may be possible without surface flow. Equation (8) should be adopted to a slope ranged 15-22°, and Fig. 2 was intended to show the applicability of Eq. (8), but not to use for the prediction of debris flow. It is true that on a steeper slope, the depth of the seepage flow should be smaller than that on a gentler slope. This means that a surface flow will more easily occur on a gentler slope than on a steeper one. On a slope steeper than 22°, the critical rainfall would be given by

$$r_T = \frac{1}{T}\int_0^T r dt \geq \frac{H_c k}{l} tan\theta$$

where, H_c is the critical depth of subsurface flow and given as

$$\frac{H_c}{D} = \frac{c/(\rho g D cos\theta) + C_*(tan\phi - tan\theta)\sigma/\rho}{tan\theta + C_*(tan\phi - tan\theta)}$$

From these two equations, it is clear that the critical rainfall decreases with slope angle as shown in the figure below.

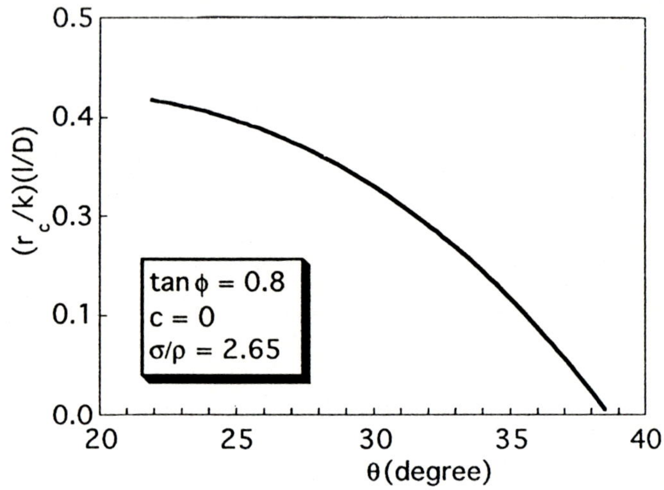

Fig. 8 Critical rainfall vs slope angle for a debris flow

The first equation can also be written in the same form as Eq. (9) as

$$R(t,T) = \int_{t-T}^{t} r(\tau)d\tau \geq \frac{H_c k}{l} tan\theta = R_c$$

Large and Small Debris Flows - Occurrence and Behaviour

Tim R. Davies

Department of Natural Resources Engineering
Lincoln University, New Zealand

Abstract

Similarities and differences of occurrence and behaviour between large and small debris flows can be explained by the differences in the depth:grain size ratio and channel characteristics. Small debris flow surges result from slope instabilities and large grain jamming in steep, narrow channels; large flows result from propagation, coalescence and amplification of surges induced in stationary slurry masses by small debris flows arriving from tributary gullies.

Grain jamming in small gullies is expected to be related to the presence of macroviscous grain collision conditions. Surge behaviour in large flows seems to be explainable using continuity and momentum conservation principles.

1. Introduction

The many reports of field debris flows in the literature suggest that there are two fairly distinct varieties of this phenomenon. In parts of China, very large debris flow surges occur during the wet season; such events can occur many times per year in the same location, and instantaneous flow rates of up to $2000m^3/s$ have been recorded. In most other mountainous areas of the world, non-volcanic debris flows are typically small (instantaneous discharges $<<$ $100m^3/s$) and infrequent, recurring perhaps once every few years in a given channel. The many other differences between these two types of flow (see Table 1) raise the question of whether these two phenomena are essentially different, in other words whether they require separate explanations.

It is the hypothesis of this paper that the properties of the flow material are essentially the same in the two cases, and that the variation in behaviour results from the different circumstances in which each occurs. Semi-mechanistic explanations are proposed for each case, based on grain collision conditions and principles of mass and momentum conservation, which appear to be in accord with field data.

2. Small Gully Debris Flows (Type A)

These are typified by the Mount Thomas, New Zealand, flows described by Pierson (1980, 1981). These occur in a small ($< 1km^2$) steep ($\geq 10\%$) catchment, every few years when intense rain falls on the already saturated catchment. Several surges occur irregularly per event; surges are $\sim$ 1 m high, move relatively slowly ($\sim$ 1 m/s) and often jerkily. Between surges, streamflow is of low density, rapid and turbulent. Surge fronts contain conspicuously more large grains ($\geq 1m$ in diameter) than the rest of the flow. Field observations outline the processes leading to such events (Davies *et al.*, 1992). Intense slope erosion causes input of fine material to the channel, forming a dense slurry and mobilising grains of all sizes. These grains assemble into large accumulations which cause build-up of slurry behind them, which eventually overtops the temporary dam or causes it to slide into motion, generating a moving surge. This sequence of events only occurs if the fine material input is sufficient to form a dense slurry ($\gamma_s \sim 1.6-1.8t/m^3$); otherwise the event will not progress beyond being a 'normal' high streamflow with coarse material moved as bedload at the base of the flow, in which no damming or surging occurs.

3. Large Valley Debris Flows (Type B)

These are typified by the Jiangjia Ravine, China, flows reported and discussed extensively in the literature (Kang and Zhang, 1980; Li, *et al.*, 1983). They occur in a large (48 km2) catchment (whose valley slope is about 7) several times per year during the south-west monsoon. Each event comprises of the order of 10-100 surges, which often occur at quite regular intervals; the surges are up to $5m$ high, and move quite steadily at up to $10m/s$. Between surges the channel contains *stationary* material of the same composition of that in the surges. The surge fronts do not appear to differ in composition from the rest of the surge.

Table 1 summarises the occurrence, characteristics and behaviour of these two types of debris flow.

TABLE 1

Characteristic	Type A	Type B
Source	Slope erosion	Stationary slurry deposit
Channel order	1	2
Channel slope S	Steep 12°	Flatter ~ 5
Channel width, w	Narrow $\sim 5\, D_{max}$	Wide $\sim 100\ D_{max}$
Max. flow depth, d	$\sim D_{max}$	$\sim 10\, D_{max}$
Channel roughness	$\sim D_{max} = d$	$\sim D_{max} = 0.1$ d
Max. grain diam. D_{max}	$\sim 1m$	~ 0.3 m
D_{75}	$\sim 0.1m$	$\sim 0.1m$
Bulk density	$1.7 - 2.2t/m^3$	$1.4 - 2.2t/m^3$
Surge period	Irregular $\sim$ 10 min	Regular $\sim$ 1 min
Motion	Slow $\sim 1m/s$, interm.	Fast $\sim 10m/s$, cont.
Between-surge flow	Turbulent streamflow	Stationary slurry
Grain jamming	Common	Absent
Appearance	Coarse, blocky head; more fluid tail	Like wet concrete (head and tail)
Frequency of events	1/year	$\sim$ 10/year

Field observations (Davies,*et al.*, 1991, 1992) show that large surges can arise from small undular surface waves in stationary slurry material; these amplify as they translate downstream, then break and continue growing in amplitude. What causes the initial small waves in the stationary material is not yet known. As the large waves move downstream, smaller and slower waves are incorporated by them and the waves become more regular, less frequent, and larger.

4. Analysis

From Table 1 it is clear that the two types of debris flow behave and appear very different(ly). It is hypothesised that this effect is mostly due to the different scales of the two situations, in particular to the ratio of surge depth to grain size.

The large, deep Type B flows typically have depth:grain size ratios of 10 or more, and have a very fluid appearance, strikingly similar to that of wet concrete; individual grains or groups of grains do not appear to influence the surge behaviour significantly, and the composition of the surge material varies little, if at all, from head to tail.

The smaller, shallower Type A flows on the other hand are usually about as deep as the largest grains, and individual grains thus significantly affect the behaviour of a surge. A single boulder at the front of a surge can slow or stop it.

This single difference, together with the fact that the first-order stream channels in which Type A flows occur are considerably smaller, steeper and narrower than the (usually) second-order streams of Type B flows, can explain all the differences listed in Table 1. The large difference in depth: grain-size ratio between the two situations is not due only to the difference in flow depths, since the maximum grain size in Type B flows is usually less than that in Type A flows, because the largest boulders ($\sim$ 1 m dia.) carried by the former are often left behind in levee-type deposits, while the rest of the grains flow on to form the stationary valley deposits that spawn Type B flows.

The difference in depth/grain size means that, while Type B flows can be analysed as fluids (see below) with some success, analysis of Type A flows requires explicit consideration of solid sliding friction of large grains and their retarding effect on the flow. Indeed, the conditions necessary for debris flows to form in steep gullies are those that allow mobilisation of large grains, in other words, a very dense slurry of fine material in water that can pick up and disperse grains of all sizes. The next section of this paper examines these conditions in more detail, while the final section explores a fluid-flow analysis of Type B surges.

5. Initiation of Type A Flows

Davies (1986) has proposed that, in debris flows, grain collision conditions are macroviscous in the sense used by Bagnold (1956), that is, that all the momentum acquired by a grain at a collision is transferred to the interstitial slurry before the next collision; such conditions imply dispersal of grains throughout the whole flow depth. Any grains eroded from the flow boundaries immediately become part of the flow, and grains of all sizes are able to be moved; since all grains in the flow add to the downslope gravity component driving the flow, there is no theoretical limit to the ability of the flow to transport grains, as long as the total solids concentration does not exceed the theoretical maximum of about 0.91 (Davies, 1988). Such flow conditions clearly allow the very largest grains present to be transported as long as $D \not> d$.

Bagnold (1955) further shows that, when shear stress in a macroviscous flow of already high concentration increases (due to an increase in flow depth, or slope, or grain concentration), this increase of shear stress causes the grain-carrying capacity of the flow to decrease. In a normal, bedload-transporting flow with inertial grain collision conditions a decrease in capacity would cause large grains to deposit on the bed; since the large grains in a macroviscous flow are uniformly dispersed, however, and in fact form an integral part of the grain-fluid mix which constitutes the flow (moving at essentially the same speed as the intergranular fluid), such deposition of large grains is impossible. The only way in which fewer large grains per unit time can be transported, therefore, is if the velocity of the flow (grains and fluid) decreases. Such a decrease will cause more rapidly-moving flow from upstream to accumulate at the section of lower velocity, increasing its depth and shear stress so that the flow capacity and velocity are further reduced. Clearly this situation can lead to complete jamming and halting of the flow.

Bagnold (1955) gives eq.(1) as the criterion for the occurrence of macro-viscous flow:

$$G^2 = \frac{\sigma(\sigma - \rho)gD^3 \cos\beta\, C_b d}{\lambda \eta_a^2} < 100 \tag{1}$$

where σ = density of solid grains; ρ = density of intergranular fluid; g = gravitational acceleration; β = bed slope; γ = linear grain concentration; C_b = volumetric grain concentration; D = grain diameter; d = flow depth and η_a = apparent dynamic viscosity of intergranular fluid. If the intergranular slurry has a density of $1.5t/m^3$, then with $d = 0.5m$, $D = 0.06m$, $\beta = 6$, $C_b = 0.3$ and $\gamma = 3$, the value of a required for macroviscous flow is $\sim$ 1 kg/m/s or about 1,000 times that of water. Laminar flow of the intergranular fluid, treated as a Bingham fluid, is also a requirement for uniform distribution of coarse grains, and occurs at a Reynolds' number R_e of about 2500 (Zhang Hao, *et al.*, 1980; Qian Yiyang, *et al.*, 1980);

$$R_e = vd\rho\eta_a \tag{2}$$

where v is the mean flow velocity. If the velocity gradient of the flow (= 3v/d, see below) is about 25, then under the above conditions $v \simeq 4m/s$ and $\eta_a = 1.2kg/m/s$. Hence, under these 'normal' flow conditions, $\eta_a \simeq 1kg/m/s$ is sufficient to cause macroviscous flow.

From tests on the rheology of a debris flow, Dai,*et al.* (1980), show how the shear strength τ_B and Bingham viscosity τ_B vary with concentration (Fig.1;

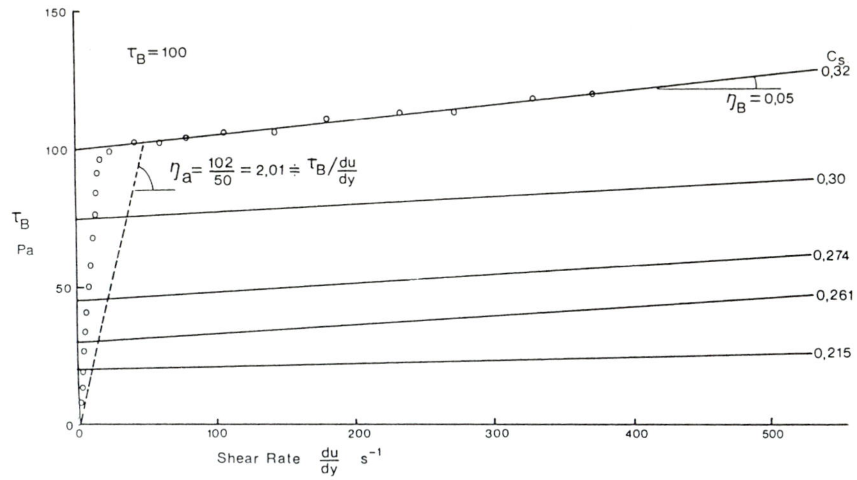

Fig. 1. Rheology of fine material slurry (Dai *et al.*, 1980)

note that τ_B is not necessarily the true yield strength of the slurry, but is obtained by back-extrapolation of the linear part of the curve to $du/dy = 0$) and it is interesting that Wan (1982) and Rickenmann (1990) find very similar results with Kaolinite slurries. The apparent viscosity a of the slurry at a given shear rate can then be found (Fig.1):

$$\eta_a = \frac{\tau}{du/dy} = \frac{\tau_B + \eta_B du/dy}{du/dy} = \eta_B + \frac{\tau_B}{du/dy} \tag{3}$$

and Fig.2 shows how η_a varies with slurry concentration C_s and du/dy.

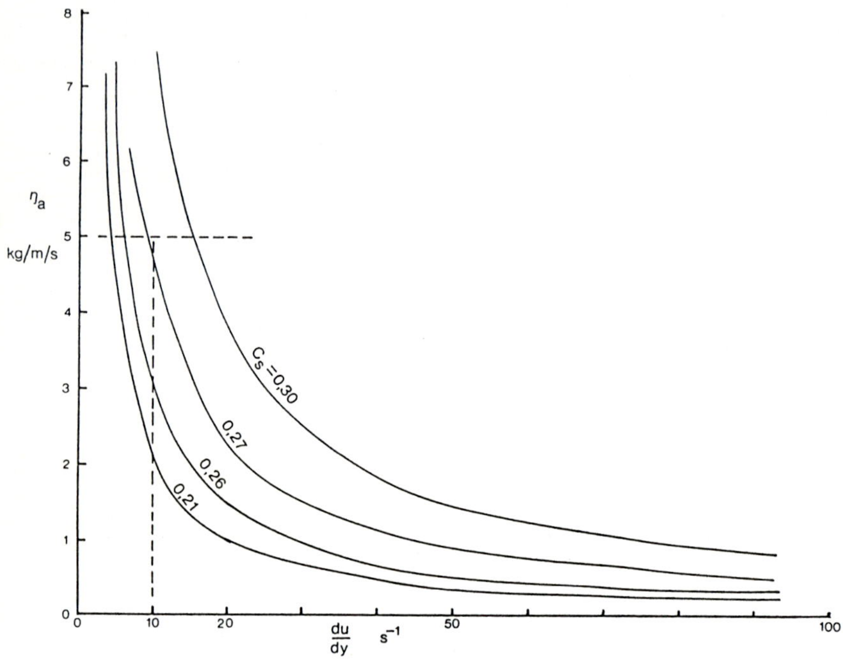

Fig. 2. Apparent viscosity a as a function of fine material slurry concentration C_s and shear rate

Various combinations of C_s and du/dy give $\eta_a = 1$. By integrating the Bingham flow equation, Rickenmann (1990) has shown that for laminar open-channel flow

$$\frac{du}{dy} = 3\frac{v}{d} \tag{4}$$

Pierson (1980) found that in the intervals between pulses at Mt Thomas, New Zealand, $du/dy \simeq 25$, which from Fig. 2 gives $C_s \simeq 0.25$ as the slurry concentration at the onset of pulsing flow, corresponding to a bulk density of $\gamma \simeq 1.41\, T/m^3$.

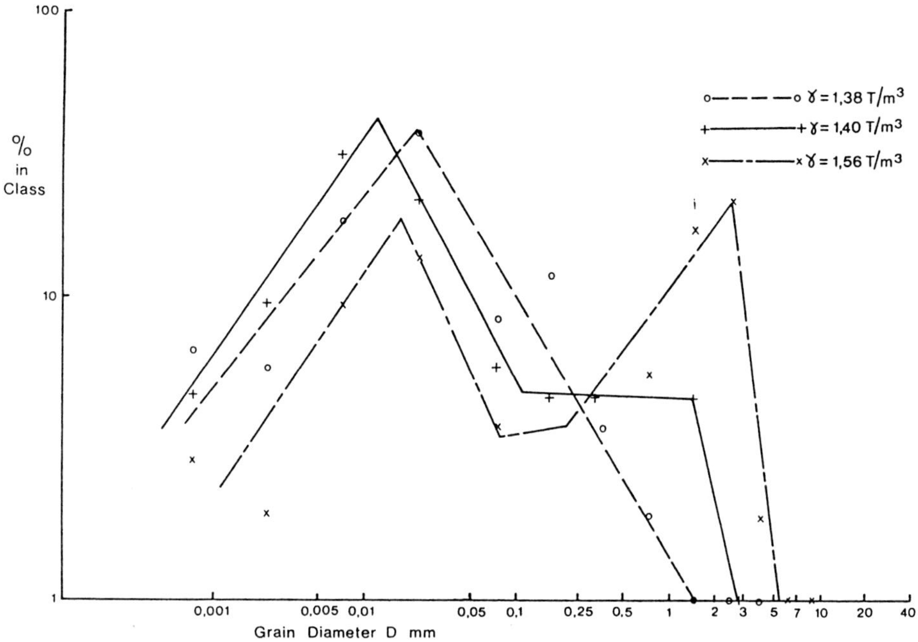

Fig. 3. Grain size distributions (non-cumulative, logarithmic scales) of debris flow material, Jiangjia Ravine, China (Li *et al.*, 1983)

The near-surface granulometry of some Chinese debris flows is shown in Fig. 3; although these are Type B flows, the material properties are almost the same as those of Type A flows and so the deductions apply to flows in small, steep gullies. It is seen that at $\gamma = 1.56$, the size distribution of grains is strongly bimodal, showing that coarse grains are distributed throughout the flow as in macroviscous flow. At $\gamma = 1.40$, there is a slight indication that coarse grains are being dispersed, whereas at $\gamma = 1.38$ the size distribution is unimodal, and no coarse grains are present at the flow surface. It appears that $\gamma = 1.40$ indicates onset of the macroviscous flow, and confirms that in this case $\eta_a \sim 1$ is a realistic criterion for predicting the conditions under which a pulsing flow will occur. Note from Fig. 2 that a considerable variation in du/dy affects this result only slightly owing to the shape of the C_s curves. Assuming, then, that for macroviscous flow $\eta_a \geq 1$, then from (3)

$$\eta_B + \frac{\tau_B}{du/dy} > 1$$

or, since $\eta_B \sim 0.01$ (Dai,*et al.*, 1980; Wan, 1982)

$$\frac{\tau_B}{du/dy} > 1$$

and giving du/dy its likely value of about 25, the criterion for the occurrence of pulsing becomes

$$\tau_b \geq 25\,Pa \tag{5}$$

Chinese data allow (5) to be tested. In Quinshui Gully, Dachao River, $\tau_B < 10Pa$ and no pulses occur, whereas in the adjacent Hunshui Gully $\tau_B \sim 30 - 50Pa$ and pulses do occur (Li and Luo, 1981). In Jiangjia Gully pulsing flows occur with $\tau_B > 200Pa$ (although there are indications that the *slurry* τ_B may be less than this), while in a different Hunshui gully $\tau_B \sim 6-20Pa$ and turbulent, non-pulsing debris flows occur which may or may not be macroviscous (Li, *et al*, 1983; Zhang, *et al.*, 1985). The simplified criterion (5) thus seems to work reasonably well, and other data support it (Pierson, 1980; Costa, 1984; Johnson and Rodine, 1984).

It thus seems reasonable to associate the onset of macroviscous flow with the presence of large grains in samples taken from the upper layers of a debris flow, indicating a high degree of grain dispersion.

In the steep, narrow gullies in which Type A flows originate there are many channel boundary nonuniformities able to disturb the 'uniform' flow of a dispersed slurry, and any of these will be able to initiate the instability which gives rise to incipient surge waves as outlined earlier. Such a surge will amplify rapidly and may well jam in a narrow channel due to the bridging of clusters of large grains across the channel, or between the bed and the free surface (see Savage and Sayed, 1989, p. 411, for a discussion of grain jamming; Bagnold (1955) also reports such an effect in an experimental flume). A blockage forms a stationary or slow-moving dam, behind which material still in motion builds up until the downstream force is sufficient to overcome the interparticle or particle-boundary friction, and the jammed material is set in motion again, moving away downstream as a surge wave. Such a surge, with a large depth and a steeply-sloping front, will exert a very high shear stress on the bed and Bagnold (1956) shows that in some circumstances the bed may be scoured by a macroviscous flow to almost unlimited depth.

Surges such as those described above will clearly occur in a more or less random sequence in small gullies, and some other mechanism is needed to explain how these surges can evolve into the much larger, more regular surge waves in the larger channels downstream described by Li, *et al.*, (1983), and Li and Luo (1981). These are up to 5 m high, 50 m wide and travel at speeds of up to 13 m/s, and it is inconceivable that they are the unmodified result of temporary blockages in small gullies, as consideration of the volume of material in a single pulse (up to $25,000m^3$) shows. This problem is examined in the following section.

6. Motion of Type B Flows

Type B flows seem to be initiated from small surface waves in stationary slurry deposits on the gently-sloping channel beds of 2nd or greater order valleys. It is supposed (in the very understandable absence of direct observational evidence, apart from that of Davies, *et al.* (1991, 1992), towards the end of an event) that Type A flows initiated in one or more first-order gullies spill out onto the channel bed of a valley, and halt there due to their viscosity, forming a stationary deposit; further Type A flows entering this deposit cause surface waves to propagate downstream. These waves break and, on reaching the shallower downstream end of the deposit, extend the deposit downstream over the rough gravel bed of the inter-event river, becoming shallower and eventually halting. Each subsequent wave extends the deposit further downstream and deepens the deposit at a given section. Waves are generated at random in the stationary deposit by entry of Type A flows; as the waves move downstream, however, the larger and faster ones overtake and incorporate the smaller and slower ones, leading to an overall increase in size, regularity and period of the waves with distance downstream (Fig. 4).

Video film of Type B surges at Jiangjia Ravine shows that a surge, on entering a deeper pool of stationary material, becomes lower, increasing in size again as it leaves the pool and enters shallower deposits downstream. This, like the general appearance of the surge, is very reminiscent of the behaviour of a moving surge in water; it suggests the hypothesis that the behaviour of debris flow surges might be amenable to analysis on the same bases, i.e., conservation of mass and momentum, as surges in water.

The analysis of a uniformly progressive wave in still water by Chow (1959, eq. 19-9) can be applied to Type B debris flow surges, since both advance into stationary fluid.

The speed of advance of the wave, Vw, is then given by

$$V_w = (g(d_o + 3h + \frac{h^2}{d_o}))^{1/2} \tag{6}$$

where d_o is the depth of stationary fluid and h the height of the surge; see Fig. 5.

That this equation might be valid for both water and debris flow material (which is about twice as dense as, and hundreds of times more viscous than,

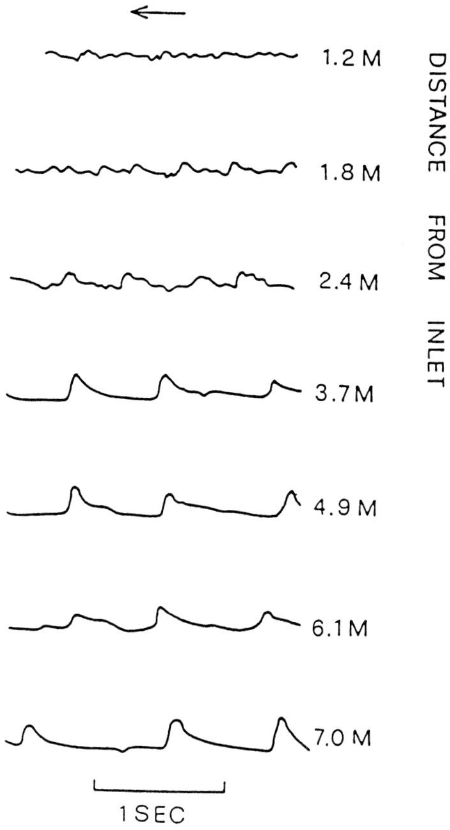

Fig. 4. Development of roll waves along a channel (Mayer, 1959)

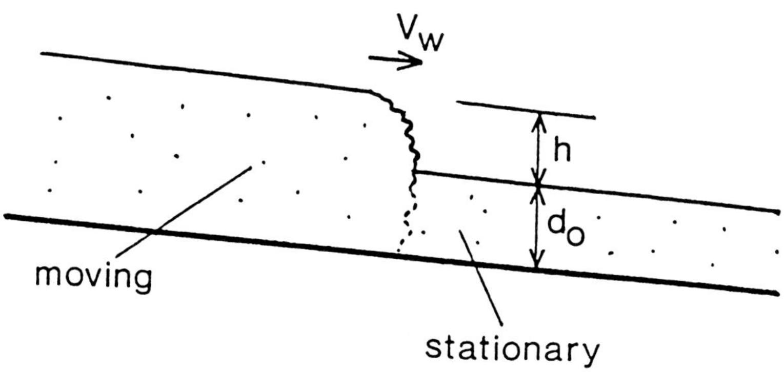

Fig. 5. Surge advancing into stationary material (Eq. 6)

water) results from density terms cancelling out and viscosity not being a factor (since energy dissipation is not considered). The equation will only be valid if the debris flow material is homogenous, i.e., the same material is found in the surge and in the preceding (stationary flow); this is not the case with Type A flows.

Data from Jiangjia Ravine (Kang, 1987) record observations of surge height and velocity in Type B flows, but unfortunately do not give information about the depth of stationary slurry preceding the surge. From other information in Kang (1987), however, this appears to be of the order of $0.5m$ in most surges, and with upper and lower bounds of 1.0m and 0m respectively. The recorded surge heights and velocities can all be reconciled with eq.(6) above with d_o between these limits. While by no means proving the validity of the equation, the data do thus indicate that it might well be valid. Based on a similar analysis, the behaviour of a surge of constant discharge moving through a deep pool of material (such as is likely to occur at channel bends (Davies, *et al.*, 1991) can be studied.

From eq. (6) it can be shown that

$$q = \left(\frac{g}{2}\frac{(d_o + h)^3(2d_o + h)}{d_o}\right)^{1/2} \tag{7}$$

where q is the discharge per unit width. With a discharge/unit width of $10m^3/s/m$, which is about the average recorded at Jiangjia, a surge wave $1.3m$ high will move at about $8m/s$ if the stationary slurry depth is $0.4m$. If this wave enters a deeper zone, of $2.0m$ depth say, the wave height will reduce to about $0.2m$ and the velocity to about $4.75m/s$, assuming that the discharge of the surge remains constant. A slight increase of d_o to $2.2m$ reduces the wave amplitude to almost zero. This would result in very low waves being apparent in the stationary slurry, which would increase in amplitude as the pool became shallower downstream; breaking would occur when $h \geq d_o$ if the fluid were water, but might occur with very much lower debris waves given the presence of quite large solid grains in the 'fluid'.

This behaviour is a possible explanation of the apparently spontaneous appearance and growth of surface waves in a stationary body of material observed by Davies, *et al.* (1991, 1992), at Jiangjia Ravine.

7. Conclusions

The dramatic differences in behaviour between large and small debris flows are explainable in terms of the different scales of the phenomena, in particular the difference in depth:grain-size ratio in the two cases; the debris flow material properties in the two cases are very similar.

The occurrence of small debris flows requires sufficient fine material in the water flow so that grain collision conditions are macroviscous; a rough criterion for this is that the shear strength of the slurry is greater than about 25 Pa.

The size and velocity of large debris flow surges can be explained by applying principles of mass and momentum conservation to a surge advancing and incorporating a stationary bed of debris flow material.

REFERENCES

Bagnold, R.A. (1955). Some flume experiments on large grains but little denser than the transporting fluid. Proc. Inst. Civil Eng., Pt 3, Paper No.6041, 174-205.

Bagnold, R.A. (1956). The flow of cohesionless grains in fluids. Phil. Trans. Royal Soc. London, 249A.

Chow, V.T. (1959). Open Channel Hydraulics. McGraw-Hill, 680 p.

Costa, J.E. (1984). Physical Geomorphology of Debris Flows. In Developments and Applications of Geomorphology, ed. J.E. Costa and P.J. Fleischer. Springer.

Davies, T.R.H. (1986). Large debris flows - a macroviscous phenomenon. Acta Mechanica. Vol.63, 161-178.

Davies, T.R.H. (1988). Debris flow surges - a laboratory investigation. Mitteilung Nr.96, VAW, ETH-Zurich, Switzerland, 96 p.

Davies, T.R.H.; Phillips, C.J.; Pearce, A.J.; Zhang, X.B. (1991). New aspects of debris flow behaviour. Proceedings, U.S.-Japan Workshop on Snow Avalanche, Landslide, Debris Flow Prediction and Control, Tsukuba, Japan.

Davies, T.R.H.; Phillips, C.J.; Pearce, A.J.; Zhang, X.B. (1992). Debris-flow behaviour - an integrated overview. Proceedings, Int. Symp. on Erosion, Debris Flow and Environment in Mountain Regions, Chengdu, China, I.A.H.S. Publication No.206, 217-226.

Johnson, A.M., and Rodine, J.R. (1984). 'Debris Flow', Ch.8 in Slope Instability, ed. D. Brunsden and D.B. Prior, J. Wiley and Sons.

Kang Zhicheng, *et al.* (1987). A comprehensive investigation and control planning for debris flow in the Xiaojiang River of Yunnan Province. Scientific and Technical Publishing Co., Chongching, Sichuan, China (in Chinese).

Kang Zhicheng and Zhang Shucheng (1980). A preliminary analysis of the characteristics of debris flow. Proc., Int. Symp. River Sed., Beijing, China, 1, 213-226.

Li Jan; Yuan Jianmo; Bi Cheng and Luo Defu (1983). The main features of the mudflow in Jiangjia Ravine. Zeit. Geomorph., 27, 3, 325-341.

Li Jan and Luo Defu (1981). The formation and characteristics of mudflow and flood. Zeit. Geomorph., 25, 4, 470-484.

Mayer, P.G. (1959). Roll waves and slug flows in open channels. J. Hydraul. Div. A.S.C.E., 85, 99-141.

Pierson, T.C. (1980). Erosion and deposition by debris flows at Mt Thomas, North Canterbury, New Zealand. Earth Surf. Proc., 5, 227-247.

Pierson, T.C. (1981). Dominant particle support mechanisms in debris flows at Mt Thomas, New Zealand, and implications for flow mobility. Sedimentology, 28, 39-60.

Qian Yiyang, *et al.* (1980). Basic characteristics of flow with hyperconcentration of sediment. Proc. Int. Symp. on River Sed., Beijing, China, 1, 175-184.

Rickenmann, D. (1990). Bedload transport capacity of slurry flows at steep slopes. Mitteilung Nr. 103, VAW, ETH-Zü rich, Switzerland, 249 p.

Savage, S.B.; Sayed, M. (1984). Stresses developed by dry cohesionless granular materials sheared in an annual shear cell. J. Fluid Mech., 142, 391-430.

Wan Zhaohui (1982). Bed material movement in hyperconcentrated flow. Series Paper 31, Inst. Hydrodyn., T.U. Denmark, Lyngby, 79 p.

Zhang Hao, *et al.* (1980). Settling of sediment and the resistance to flow at hyperconcentrations. Proc. Int. Symp. on River Sed., Beijing, China, 1, 185-194.

Zhang Xinbao, *et al.* (1985). The main features of debris flows and control structures in Hunshui Gully, Gunnan Province, China. Proc. Int. Symp. on Erosion, Debris Flow and Disaster Prevention, Tsukuba, Japan, 181-186.

DISCUSSION

Aguirre Pe: The diameter D that you have considered in Type A Flows refer only to larger particles or also to particles in the interstitial spaces between larger stone?

Do interstitial particles include small sand particles, clay and silt?

Davies: I assume that the debris flow consists of coarse grains ($D > 1 - 5mm$) in a fluid slurry. Grains finer than $\sim 1mm$ make up the slurry, with water.

The diameter D in Eq. 1 is somehow representative of the coarse grains, and is in the range $5mm - 500mm$. All the numerical quantities referredto are really only orders of magnitude - the analyses are not supposed to be exact at all.

Armanini: In many European countries, debris flows cannot be included in Type A or Type B, but they have the general geometrical characteristics of Type B. In fact, often debris flows occur on steep channels as in your Type A, large boulders are present but the motion appears continuous and the speed of the flow is high like in your Type B. The sediment concentration is relatively high ($C > 0.3$), but the amount of fine particles (clay) is scarce. In which category could such a kind of debris flow be included?

Davies: Type B flows also have high grain concentration - up to 60% or γ is $2.2t/m^3$. It is to be expected that flows occur with the characterstics result from boundary conditions rather than from differences of internal processes. Stony debris flows such as those experienced in Italy might well be Type A, with low viscosity interstitial fluid.

Armanini: A comment on wall roughness effect on debris flow. In laminar flow we do not expect that wall roughness affect the flow resistance. In turbulent debris flows solid particles tend to occupy the spaces among the elements of wall roughness, so that to make the wall smoother. When the concentration is high, the particle collisions become much more important for the flow resistance than wall roughness. In Fig. 6 is reported the Strickler coefficient for a debris flow as a function of particle concentration, measured by Armanini and Scotton [1993] in experimental investigations.

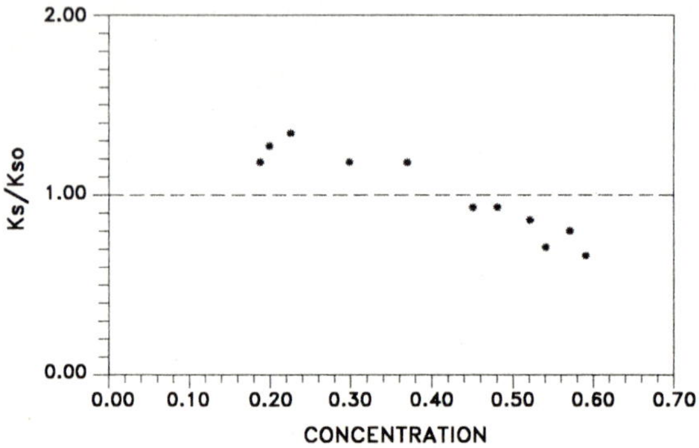

Fig. 6 k_s versus concentration. Slope = 25% (Scotton and Armanini (1993), Effects of bed roughness on debris flow, Proceedings of XXV Congress of IAHR, Vol. 3, Tokyo pp. 63-70)

Davies: As long as the flow does slide along a smooth boundary, I agree. Laminar flow is not affected by boundary roughness.

If the boundary is very rough, so that boundary roughness ¿¿grain diameter, then the boundary can perhaps affect the internal flow. This situation is unlikely in practice, since the boundary is probably made from deposits of previous flows.

Debris flow is divided into two cases by the grain size ration in your report (p. 17). But I think that this difference is due to a resistance for the flow depth.

Davies: The channel cross-section, on width/depth ratio, will affect the flow velocity for give γ, slope and depth. So flows in the deep, narrow Tsagiu gully will flow more slowly than in main Jiangjia ravine, but also I think Tsagiu flows are usually denser than those in Jiangjia.

Taniguchi: 1. How do you estimate the value of η_a in a real debris flow?

2. Do you think that the most important factor is the size of composing material of a debris flow?

Davies: 1. If we know the depth d, and we know the surface velocity v and density γ, then $\gamma = \gamma d S$, S=surface slope, $dv/dy = 3v/d$, $\eta_a = \tau/(dv/dy)$.

2. I think that in Type A flows, the front of the flow controls the behaviour of the flow. The front has high grain concentration, high grain friction, and the largest grains-the size of grains therefore is a very important factor.

Field Survey for Debris Flow in Volcanic Area

Etsuro Shimokawa and Takashi Jitousono

Department of Hydraulics and Ocean Engineering
Kagoshima University
Kagoshima 890, Japan

Abstract

In this paper, the hydrological and geomorphological characteristics of debris flow associated with volcanic eruptions were examined mainly on the basis of field observations and measurements at the three volcanoes, Sakurajima and Unzen volcano in Japan and also Merapi volcano in Indonesia.

Covering or deposition of tephra from ash eruptions or pyroclastic flows radically altered the hydrological and erosion regime of basins in the three volcanoes. As the results of those the debris flows which originate in the sheet-rill-gully erosion induced by surface runoff, frequently occurred even under less rainfall. The critical rainfall which is defined by the two parameters, the rainfall just before the occurrence of debris flow and the cumulative rainfall from its beginning, was considerably low for several years after as well as during the volcanic eruption. The analysis of runoff characteristics shows that a most part of the debris flows in the volcanic areas are of muddy type including much fine pyroclastic materials. The average sediment concentration obtained from a relationship between the total runoff and the total sediment yield is available for evaluating the sediment yield by a debris flow or debris flows at a rain.

1. Introduction

Many field surveys have been done in regard to debris flow until now. They involve geomorphological, hydrological and hydraulic observations and measurements. The morphological features and mechanism of each process of the initiation, flow and deposition have been made clear through the field surveys. In particular, direct observation of the debris flow by motion picture photography which firstly started in the 1970s in Yakedake volcano (Suwa et al. 1973) and then was followed in Sakurajima volcano (Watanabe and Ikeya

1981, Haruyama et al. 1984, Jitousono and Shimokawa 1991), contributed much to elucidating the motion dynamics and runoff characteristics of debris flow.

The purpose of this paper is to examine the hydrological and geomorphological charactreritics of debris flow associated with tephra from volcanic eruptions through the field surveys at the three volcanoes, Sakurajima, Unzen and Merapi volcano.

2. An Outline of Study Areas

Sakurajima situated at the northern part of Kagoshima bay, southern part of Kyushu, Japan, is a composite stratovolcano with two peaks of 1117 and 1011 m above sea level. It is one of most active volcanoes in Japan and has been active with frequent ash eruptions over a long period since 1972. The debris flows associated with the volcanic activity have occurred many times every year along each river on the flank.

Unzen volcano located at the Shimabara peninsula, northwestern part of Kyushu, is a composite stratovolcano consisting mainly of dacitic rocks. It started erupting with ashfall in November 1990 after 198 years of dormancy since the last eruption in 1792 and has erupted with the pyroclastic flows due to collapsing lava dome on the peak of Mt. Fugen since May 1991. Following the ash eruptions and pyroclastic flows the debris flows have frequently occurred along the valleys filled by the pyroclastic flow deposits.

Merapi volcano whose peak is 2947 m, is a composite stratovolcano locating in the central part of Java island. The volcano which is one of the most active volcanoes in Indonesia, has often erupted with pyroclastic flows induced by collapsing lava mass. The pyroclastic flows are followed by lahar or volcanic debris flows. Recently, larger scale pyroclastic flows occurred in June

1984. The pyroclastic flow deposits widely covered the upper reaches of the Putih and Bebeng rivers situated at the southwestern flank of the volcano.

3. Rapid Alteration of Hillslope Hydrology by Tephra Covering

Covering or deposition of much tephra produced by volcanic eruptions is generally followed by a radical alteration of hydrological regime and then by intense erosion and much sediment yield on the hillslopes even under less rainfall and also by frequent occurrences of the debris flow in basins, as shown in Fig.1.

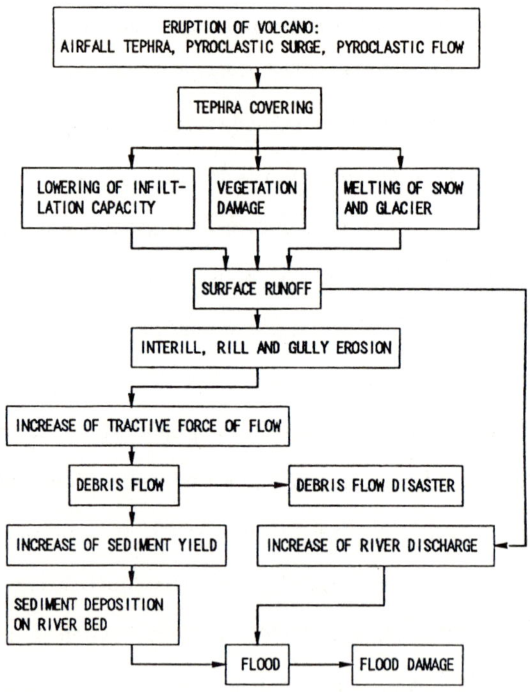

Fig. 1. A flow chart indicating an effect of tephra covering on occurrence of debris flow and flood

The hydrological surveys which are aimed at investigating the effects of tephra depositions on the hillslope hydrology, involve mainly measurements of infiltration capacity and observations of surface runoff.

The hydrological observation equipments with a rain gauge were set up at the zero-order tiny catchments on the northern flank of Sakurajima (Jitousono and Shimokawa 1987, 1989b). The water-stage of surface runoff was measured by the V-notch weir with water-stage recorder installed at the lower end of the catchments and then the records were converted to the discharge. The effect of volcanic ash cover on the occurrence of surface runoff was analyzed using the records (Jitousono and Shimokawa 1987). Fig.2 shows a example of hydrographs of the surface runoff obtained at the two catchments vegetated with blackpine and broadleaved trees, respectively. The occurrence of surface runoff and its peak discharge may depend on rainfall, infiltration capacity and vegetation. A critical rainfall defined by two parameters, a rainfall intensity just before the occurrence of surface runoff and a cumulative rainfall preceding the rainfall intensity, was examined on the basis of the observation records. The rainfall intensity just before the occurrence of surface runoff is considerably low with 2 mm in 10-minute rainfall at the catchment vegetated with blackpine trees and 1.5 mm at the catchment with broadleaved trees in correspondence to the preceding cumulative rainfall of more than 20 mm in a 24-hour rainfall. The critical rainfall is harmonious with for the occurrence of debris flow (Jitousono and Shimokawa 1989a). This may be created by rapid lowering of the infiltration capacity on the hillslope associated with accumulation of volcanic ash from the frequent eruptions since 1972 in Sakurajima. The infiltration capacity during the frequent ash eruptions is considerably less than exceeding 100 mm/hr before the eruptions, ranging from 14.9 to 52.9 mm/hr. The rates just after covering of new ash were furthermore small with a range of 9.0-24.2mm/hr (Shimokawa and Jitousono 1987c).

Following the eruptions with ejection of volcanic ash from November 1990 to May 1991 in Unzen, pyroclastic flows induced by collapse of lava domes have occurred since 24 May 1991. Much fine airfall tephra from the hot clouds covered the hillslopes and created lowering of the infiltration capacity. The infiltration capacity was measured at some sites covered with the fine tephra of 10-35 mm in thickness ranges 12 to 35 mm/hr, being about one-fifth of 110-148 mm/hr prior to the eruption.

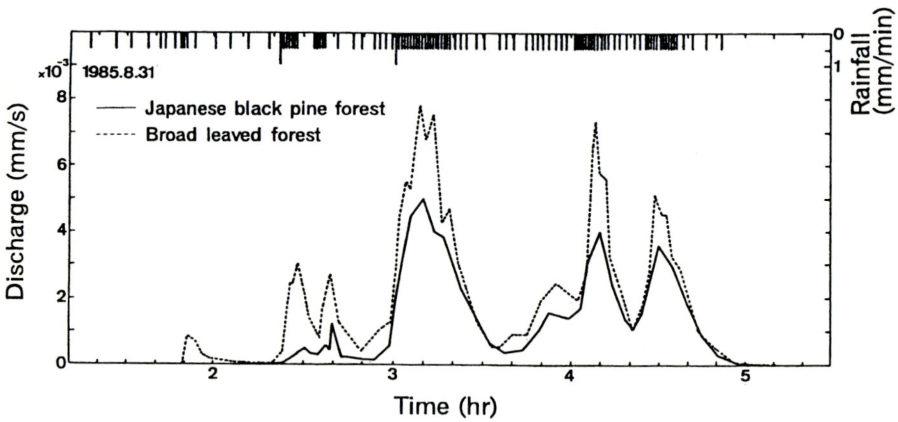

Fig. 2. A typical example of hydrographs of surface runoff at a catchment vegetated with blackpine trees and a catchment with broadleaved trees, Sakurajima (after Jitousono and Shimokawa 1987)

The lowering of infiltration capacity is confirmed in Usu volcano (Yamamoto 1984) and Mt. St. Helens, USA (Swanson et al. 1983, Janda et al. 1984a,b) as well. In Usu, fine tephra brought about a rapid lowering of the infiltration capacity with 10-minute intensity of 2 to 3 mm (Yamamoto 1984). In Mt. St. Helens, the infiltration capacity was radically reduced from approximately 100 mm/hr prior to the 1980 eruptions to 1-4 mm/hr in July 1980 of the two months later from the eruption and 7-9 and 13 mm/hr in August 1981 (Swanson et al. 1983, Janda et al. 1984a,b).

4. Sheet, Rill and Gully Erosion as Source of Debris Flow

The debris flow may originate in sheet-rill-gully erosion in a devastated volcanic area covered with new tephra. The debris flow may be initiated as follows: The surface runoff flow created from rill and gully channels increases in both discharge and fluid pressure by coming down together and by picking up debris including large boulders from the steep river bed towards the lower reaches and turns to the debris flow (Suwa et al. 1989).

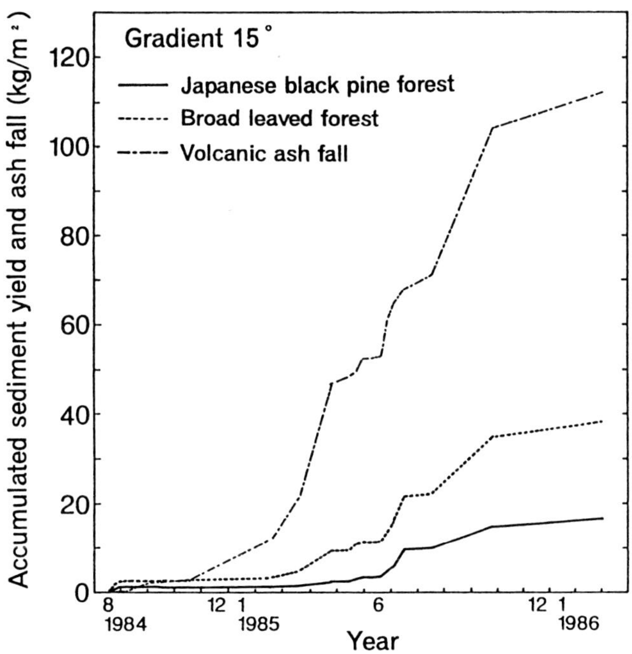

Fig. 3. A temporal variation of sediment yield by interrill erosion and of volcanic ash fall at 10 m^2 experimental plots, Sakurajima (after Shimokawa and Jitousono 1987a)

The processes of erosion and sediment yield during and after volcanic eruptions which are related to the frequency and magnitude of debris flow and their temporal variations, were examined at the three volcanoes, Sakurajima, Unzen and Merapi. Fig.3 shows a temporal variation of the accumulated sediment yield by sheet erosion at the 10 m^2 survey station with the accumulated amount of ashfall starting from August 1984 in Sakurajima (Shimokawa and Jitousono 1987a).

The accumulated sediment yield at the station increases in proportion to the accumulated amount of ash-fall. Fig.4 shows a typical example of the temporal variation of gully cross-section at the survey station (Shimokawa and Jitousono 1987b).

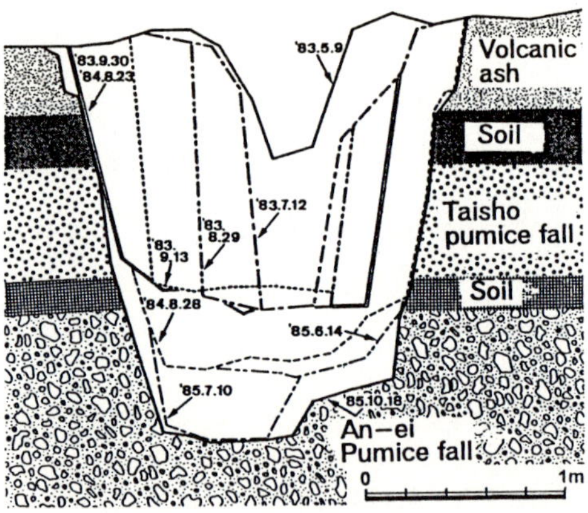

Fig. 4. A typical example of temporal variation in a cross-profile of gully, Sakurajima (after Shimokawa and Jitousono 1987b)

The cross-section has unequally developed by undercutting sometimes and by lateral expanding sometimes because the resistance to erosion varies with the geological constituents of gully bed and side wall. Mean annual sediment production per 1 km^2 by the interrill, rill and gully erosion on the hillslopes of Sakurajima is not less than 93000 m^3, coming to over 93 mm in annual erosion rate (Shimokawa and Jitousono 1987c). Judging from the previous works in Usu volcano (Chinen 1986) and Mt.St.Helens (Collins and Dunne 1986), it seems that a radical rise of the sediment yield by sheet-rill-gully erosion on the hillslopes thinly overlaid with pyroclastic airfall and surge deposits was nearly finished within a year to 2 years after the completion of volcanic eruptions because of washing out of fine pyroclastic materials by sheet erosion and of the recovery of infiltration rate. However, the high erosion rate have been kept over 20 years under a long term volcanic activity with frequent ash eruptions since 1972 in Sakurajima.

Merapi volcano located in the central part of Java island, erupted with large scale pyroclastic flows in June 1984. The pyroclastic flow deposits of approximately 6.5 million cubic meters in volume, widely covered the south-western part of the volcano and created a radical alteration of the hydrological and erosion regime of the two basins, Putih and Bebeng. As a result of that much sediment was produced by sheet-rill-gully erosion from the hillslopes and was transported by volcanic debris /mud flows and floods to the lower reaches of the two rivers. A total amount of sediment yield by sheet-rill-gully erosion from the 1984 pyroclastic flow deposits covered hillslopes was evaluated and the rates of sediment yield were estimated based on the field survey and interpretation of aerial photographs. Fig.5 shows temporal variations of the sediment yield and specific sediment yield with period rainfall at the two basins (Shimokawa et al. 1995). The sediment yield at the basins rapidly increased to a peak for a year from November 1984 to October 1985 and then slowly declined with time keeping high level for approximately 4 years until October 1988. This is harmonious with that the debris flows occurred many times for a period of approximately 4 years after the 1984 pyroclastic flows. Comparing with the effect of airfall tephra, it is assumed that the pyroclastic flows in Merapi had the effect on the sediment yield over a longer period, probably because the deposits prone to erosion filled drainage channels and the recovery of infiltration rate was retarded.

5. Hydrological and Hydraulic Surveys on Debris Flow

Video camera and ultrasonic waves water gauge are usually used for investigating the hydrological and hydraulic characteristics of debris flow. Dynamic picture photography by use of video camera is a most common method of field survey for the debris flow. In this system, arrival of a debris flow is detected by cutting of the wire sensors set across the river and simultaneously video camera and video tape recorder start recording motion picture of the debris flow. Many debris flow motion pictures have been obtained in Yakedake (Okuda et al. 1980), Sakurajima (Watanabe and Ikeya 1982, Haruyama et al. 1984, Jitousono and Shimokawa 1989a) and St.Helens (Pierson 1986). These motion pictures are available for analyzing geometrical, hydrological and hydraulic characteristics of the debris flows.

The ultrasonic waves water gauge is an uncontact type apparatus in which water stage is detected by a round time of ultrasonic waves sent from a transmitter and receiver to water surface. This apparatus is effective for observing the debris flow with much sediment as well as flood and is possible to get the records of all the debris flows and floods including small scale one throughout the year. The data are available for analyzing runoff characteristics of debris flow, water balance and sediment yield in a basin. The apparatus was

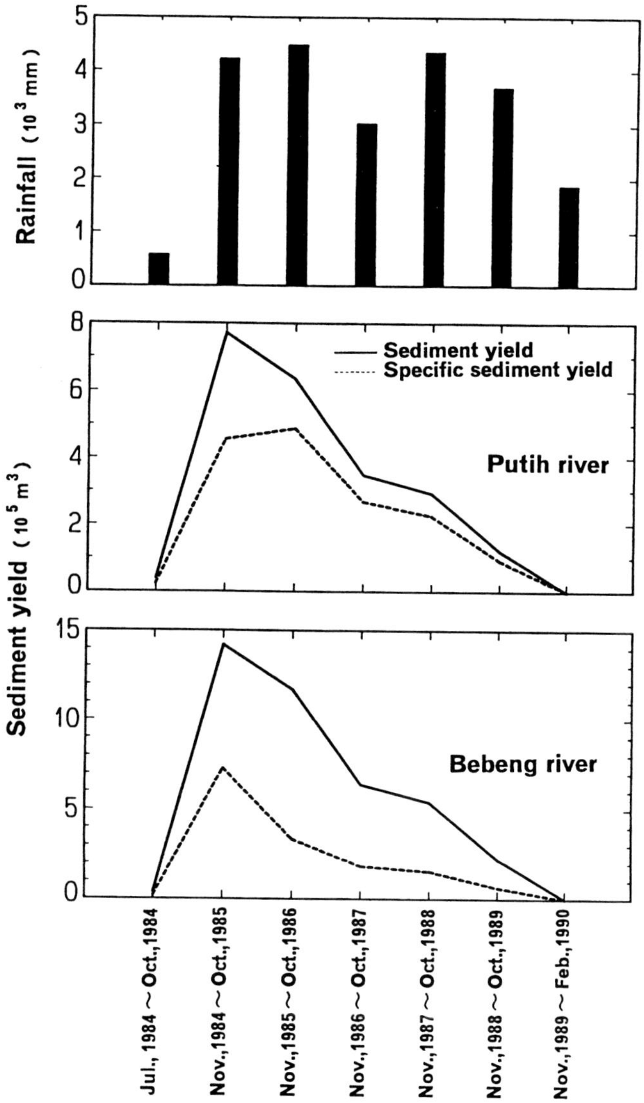

Fig. 5. Temporal variations of 4-month and annual sediment yield and specific sediment yield with 4-month and annual rainfall in Merapi volcano, Indonesia (after Shimokawa et al. 1995)

installed at each site of the lowest reaches of Saido-gawa and Arimura-gawa, Sakurajima volcano just near which video camera for observing the debris flow had been set before. All the rivers including these rivers developed radially from the summits towards the surroundings, have running water only at rainfall. The apparatus is going on operation throughout the year.

Fig.6 shows the mean monthly occurrence-number of debris flows and floods observed by the video camera and the ultrasonic waves water gauge with mean monthly rainfall for a 13-year from 1981 to 1993 at the Saido-gawa station, Sakurajima (Jitousono et al. 1995). Approximately 80 per cent of all the debris flows and floods were observed during the summer season from May to September with much monthly rainfall. The distinction between debris flow and flood depends on magnitude of peak discharge and shape of hydrograph. Using these records including rainfall data, the critical rainfall which is defined by two parameters, a rainfall just before the occurrence of debris and cumulative rainfall from its beginning, and the effect of volcanic activity on its temporal variation were examined. The rainfalls just before the occurrence of debris flow was 4 to 5, 3 to 4 and 2 to 3 mm in 10-minute intensity in correspondence to the cumulative rainfall of approximately 10, 20 and 30 mm, respectively. And the critical rainfall showed a little variation corresponding the rise and fall of volcanic activity within 13 years from 1981 to 1993. The occurrence condition of debris flow due to rainfall was investigated in Merapi volcano, also. The critical rainfall was as low as Sakurajima for approximately 4 years after the 1984 pyroclastic flows and then increased with year (Jitousono et al. 1995).

Runoff characteristics of the debris flows and floods were analyzed by use of the water-stage observation records at the two volcanoes, Sakurajima and Merapi. Fig. 7 shows a relationship between peak discharge and total runoff of the debris flows and floods at the two hydrological stations. The catchment

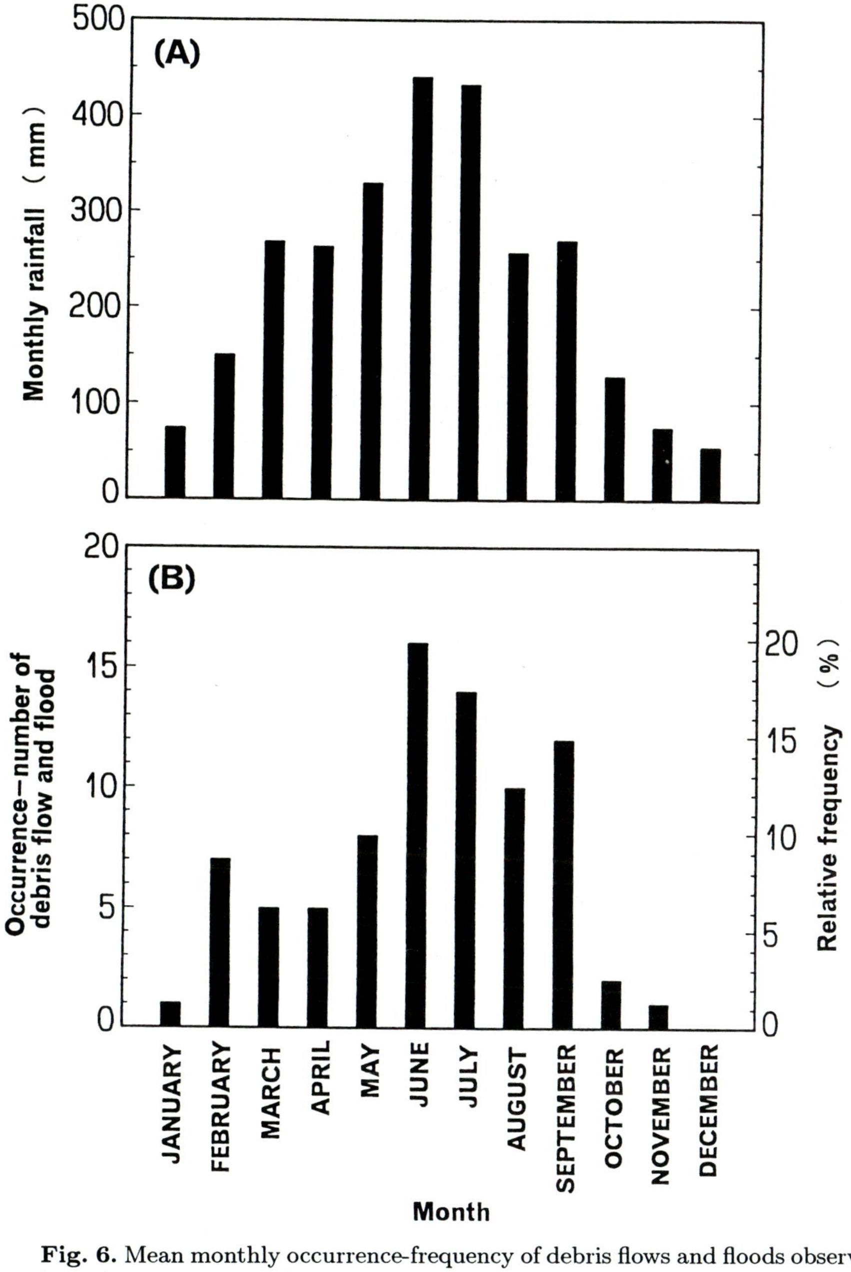

Fig. 6. Mean monthly occurrence-frequency of debris flows and floods observed by video camera and ultrasonic waves water-gauge during a 13-year from 1981 to 1993 at Saido-gawa station, Sakurajima (Jitousono et al. 1995)

area at the observation station is 1.38 km^2 in Sakurajima and 8,22 km^2 in Merapi. A nearly linear relation is obtainable on a logarithmic graph paper with three parallel alignments each other depending on catchment area and observation period. According to the observation by the video camera at the Sakurajima, the debris flows are mostly muddy type including much volcanic ash. So, the most part of debris flows at the Merapi may be mudflow.

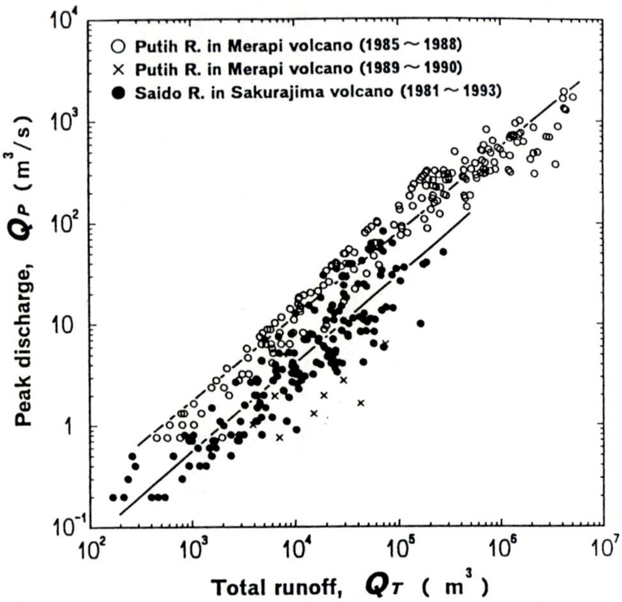

Fig. 7. A relationship between the total runoff and the peak discharge analyzed on the basis of records observed by ultrasonic waves water-gauge at Saido-gawa station (Jitousono et al. 1995)

6. Measurements of Sediment Concentration and Evaluation of Sediment Yield

Sediment concentration of a debris flow is an important factor to evaluate sediment yield as well as to examine flow behavior. The sediment concentration of debris flow, however, has not been measured so many times in fields because the sediment sampling work involves a risk and technical difficulty (Waldron 1967, Pierson 1986). A hand-powdered equipment for dip sampling the slurries was suspended over the center of a cable crossing the channel by pulley at the same site as the ultrasonic waves water-gauge, the lowest reaches of Saido-gawa, Sakurajima. Sampling of the slurry of debris flow was done many times by use of this equipment corresponding to the water-stage hydrograph of a series of debris flow. Fig.8 shows a relationship between the sediment concentration and the discharge on a logarithmic graph paper for the three debris flows. The sediment concentration linearly increases with the discharge with a extent of scatter from the regression line. The relationship, however, is only for small scale debris flows and floods with low sediment concentration of under 10 percent because the larger scale debris rarely occurred and as a result the slurry samples could not be collected.

The annual sediment yield by both of the debris flows of 18 in number and the floods of 17 observed at the Saido-gawa basin in 1992 was evaluated from both of the hydrographs and the sediment concentration at each stage of a hydrograph. The annual sediment yield from the basin of 1.43 km^2 in area is evaluated to be 101674 m^3, coming to 71100 m^3/km^2/year in specific sediment yield. In this calculation, the high sediment concentration for the larger scale debris flows was obtained by extrapolation.

The sediment concentration of debris flow which indicates an average concentration, is obtained from a relationship between the total runoff including sediment and the total sediment runoff of a debris flow or debris flows by a rain, as shown in Fig. 9 (Jitousono et al. 1995). The total sediment yield is measured from the deposits of debris flows in check dams and/or on alluvial fans. It is available only for evaluating the total sediment yield by debris flows triggered by a rain.

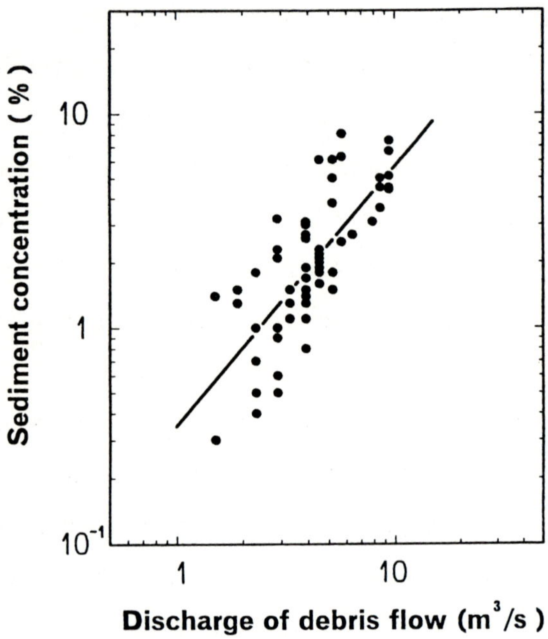

Fig. 8. A relationship between sediment concentration and discharge on a logarithmic graph paper for three debris flows at Saido-gawa station, Sakurajima

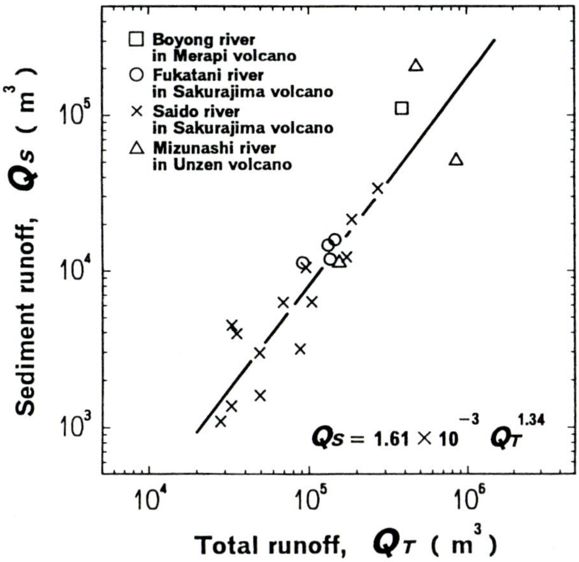

Fig. 9. A relationship between total runoff and sediment yield of debris flows induced by a rain (Jitousono et al. 1995)

7. Conclusions

The hydrological and geomorphological characteristics of debris flow associated with volcanic eruptions were examined mainly on the basis of field observations and measurements at the three volcanoes, Sakurajima and Unzen and Merapi. The results are summarized as follows:

1) The infiltration capacity largely lowered from over 100 mm/hr before the eruptions to a range of approximately one-third to one-tenth in Sakurajima and Unzen. This generated surface runoff on the tephra covered hillslopes even under a less rain with 1.5-2.0 mm in 10-minute rainfall intensity just before its occurrence corresponding the preceding 24-hour rainfall of over 20 mm in Sakurajima.

2) During and for a period after the volcanic eruptions much sediment was produced by intense erosion from the tephra covered hillslopes with some

variations depending on the type and magnitude of volcanic activity. The sediment yield from the 1984 pyroclastic flow deposits covered hillslopes in Merapi kept high rates for approximately 4 years after the 1984 eruptions.

3) The critical rainfalls are considerably low for several years after as well as during the volcanic eruptions, showing temporal variations with the rise and fall of volcanic activity and the time erapsed from the completion of volcanic eruption.

4) From the runoff characteristics of debris flow, it is assumed that the most part of debris flows are muddy type including much fine pyroclastic materials.

5) The average sediment concentration of debris flows which was obtained from the relationship between the total runoff including sediment and the total sediment runoff, is available for evaluating the sediment yield by a debris flow or debris flows occurred at a rain.

REFERENCES

Chinen, T.(1986) Surface erosion associated with tephra deposition on Mt. Usu and other volcanoes: Environ. Sci. Hokkaido, 9(1)C137-149.

Collins, B. D. and Dunne, T. (1986) Erosion of tephra from the 1980 eruption of Mount St. Helens: Geol. Soc. Ame. Bull., 97C896-905.

Haruyama, M., Jitousono, T. and Jomoto (1984) Analyses on mud flows in Sakurajima volcano: Jour. Jap. Soc. Erosion Control Engineering, 37(2), 22-27 (in Japanese with English abstract).

Janda, R. J., Meyer, D. F. and Childers, D. (1984a) Sedimentation and geomorphic changes during and following the 1980-1983 eruptions of Mount St. Helens: Washington(1): Jour. Jap. Soc. Erosion Control Engineering, 37(2), 10-21.

Janda, R. J., Meyer, D. F. and Childers, D. (1984b) Sedimentation and geomorphic changes during and following the 1980-1983 eruptions of Mount St. Helens, Washington(2): Jour. Jap. Soc. Erosion Control Engineering, 37(3), 5-19.

Jitousono, T. and Shimokawa, E. (1987) Surface runoff features on hillside slopes covered with volcanic ash in Sakurajima Volcano: Bull. Kagoshima Univ. Forests, 15, 51-61 (in Japanese with English abstract).

Jitousono, T. and Shimokawa, E. (1989a) Debris flow in northern flank of Sakurajima volcano: Proc. Int. Symp. on Erosion and Volcanic Debris Flow Technology, Yogyakarta, Indonesia, July-August 1989, V24.1-20.

Jitousono, T. and Shimokawa, E. (1989b) Surface runoff on tephra-covered hillslope in Sakurajima volcano: Jour. Jap. Soc. Erosion Control Engineering, 42(3), 18-23 (in Japanese).

Jitousono, T. and Shimokawa, E. (1991) Effects of volcanic activity on occurrence and runoff of debris flow in Sakurajima volcano: Jour. Jap. Soc. Erosion Control Engineering, 43(6), 9-15 (in Japanese with English abstract).

Jitousono, T. and Shimokawa, E. and Tsuchiya, S. (1995) Debris flow following the 1984 eruption with pyroclastic flows in Merapi volcano: Jour. Jap. Soc. Erosion Control Engineering, in print.

Okuda, S., Suwa, H., Okunishi, K. Yokoyama, K. and Nakano, M. (1980) Observation on the motion of a debris flow and its geomorphological effects: Zeit. Geomorph. N. F., Suppl., Bd.35, 142-163.

Pierson, T. C. (1986) Flow behavior of channelized debris flows, Mount St. Helens, Washington: In Abrahams, A. D. ed., Hillslope Processes, Boston, Allen & Unwin, 269-296.

Shimokawa, E. and Jitousono, T. (1987a) Sediment yield by sheet erosion from hillslopes of Sakurajima volcano: Jour. Jap. Soc. Erosion Control Engineering, 39(6), 11-17 (in Japanese).

Shimokawa, E. and Jitousono, T. (1987b) Sediment yield by rill and gully erosion from hillslopes of Sakurajima volcano: Jour. Jap. Soc. Erosion Control Engineering, 40(1), 19-24 (in Japanese).

Shimokawa, E. and Jitousono, T. (1987c) Rate of erosion on tephra-covered slopes of volcanoes: Trans. Jap. Geomorph. Union, 8, 269-286 (in Japanese with English abstract).

Shimokawa, E. and Jitousono, T. and Tsuchiya, S. (1995) Sediment yield from the 1984 pyroclastic flow deposits covered hillslopes in Merapi volcano: Jour. Jap. Soc. Erosion Control Engineering, in print

Suwa, H., Okuda, S. and Yokoyama, K. (1973) Observation system on rocky mudflow: Bull. Disast. Prev. Res. Inst. Kyoto Univ., 23, 59-73.

Suwa, H., Manaka, T. and Inaniwa, A. (1989) Occurrence of debris flows and their scales in the Kamikamihori valley of Mount Yakedake: Ann. Disast. Prev. Res. Inst. Kyoto Univ., 32B-1, 229-247 (in Japanese with English abstract).

Swanson, F. J., Collins, B. D. and Dunne, T. (1983) Erosion of tephra from hillslopes near Mt. St. Helens and other volcanoes: Proc. Symp. Erosion Control in Volcanic areas, Seattle, July 1982: Ibaraki, Japan, Public Works Research Institute, 183-221.

Waldron, H. H. (1967) Debris flow and erosion control problems caused by the ash eruptions of Irazu Volcano, Costa Rica: U.S. Geological Survey Bull. 1241-I, 11-37.

Watanabe, M. and Ikeya, H. (1981) Investigation systems and analysis on volcanic mudflow in Mt. Sakurajima, Japan: Proc. Int. Symp. Erosion and Sediment Transport Measurement, June 1981, Florence, Italy, 1-24.

Yamamoto, H. (1984) Erosion of the 1977-78 tephra layers on a slope of Usu volcano, Hokkaido: Trans. Jap. Geomorph. Union, 5, 111-124(in Japanese with English abstract).

Chapter 2

Dynamics of Debris Flow

Introduction

Pierre Julien

It is a privilege to introduce the reader to four papers on debris flows and hyperconcentrations of sediment. The purpose of this brief report is to guide the reader and to outline the most relevant aspects of the recent scientific contributions to the field of debris flow dynamics. This set of papers nicely contributes to recent developments in terms of rheology, laboratory experiments, and field verification of numerical models for the simulation of mud flows and debris flows. It is interesting to note that the dynamics of debris flows can only be captured through clear understanding of the rheology of hyperconcentrations of sediments. The reader must overcome complexities inherent to different nomenclatures and the tendency for each author to present different rheological models. This set of papers presents a complete description of the various shear stress components due to the bonding between cohesive particles, fluid viscosity including viscous interactions with sediment particles, turbulence, and dispersive stress due to inertial collisions between particles. Jan and Shen clearly present an unprejudiced review of several models with primary results summarized in five tables.

Recent advances in rheology include quadratic formulations of shear stress. The quadratic shear stress equation of O'Brien and Julien (1985) combines yield strength, viscosity, turbulence, and dispersive stress. Equivalent quadratic shear stress relationships are also found in Takahashi, and Jan and Shen. Julien and O'Brien show numerical solutions after the friction slope is subdivided into three components; the yield slope, the viscous slope and the turbulent-dispersive slope. The approach is quite simple compared to the untractable analytical solutions for velocity and sediment concentration profiles. The quadratic model includes the inertial formulation of Bagnold's dispersive stress for which experimental data has been collected in recent years for comparison with the original experiments. There is growing evidence that the dispersive stress concept is not as simple as initially pictured by Bagnold:

a) Takahashi clearly demonstrates in his Figure 2 that the coefficient f of Bagnold's equation varies by at least an order of magnitude when compared with the experiments of Daido et al. (1984) and Campbell and Brennen (1985). Commemorating the 40th anniversary of Bagnold's contribution, the reader would have expected the empirical calibration coefficient to be known with two significant digits. The large scatter in Figures 2 and 5 is rather unconvincing, considering that several laboratory experiments were deliberately carried out under conditions similar to Bagnold's original study.

b) The use of neutrally buoyant material ($\sigma = \rho$) also poses mathematical difficulties of the type $0 \neq 0$ in Equations 21 and 27 when the granular material is under deformation $du/dz \neq 0$.

The laboratory measurements of velocity profiles by Takahashi in the inertial regime (Figure 6), and by Hashimoto for dry sand (Figure 6) and sand-water mixtures (Figure 12) are particularly enlightening:

a) In all cases, the reader will notice that the velocity increases almost linearly with depth. The similarities with the model of Duboys (1879) cannot be overlooked. This has a considerable practical meaning in that despite the diverse velocity profiles suggested in the literature, the practitioner can simply use the linear velocity profile as a first approximation. Accordingly, the surface velocity is approximately twice the mean debris flow velocity. The reader should compare the velocity profiles suggested by Jan and Shen (in Tables I-V) with the observations of Hashimoto (Figures 6 and 12) and Takahashi (Figure 6). The main reason for the discrepancies is that the suggested velocity profiles are calculated assuming a uniform sediment concentration. In reality, the increased near-bed sediment concentration reduces the velocity in the lower part of the velocity profile.

b) The reader should pay attention to the average rate of deformation in velocity profiles. For instance, in Takahashi's Figure 6, $du/dz \cong 10/s$, which is very small compared to the deformation rates required for inertial particle impact in Bagnold's experiments ($50 < du/dz < 300$). Considering near-linear velocity profiles, the practitioner will notice that given a typical flow depth of 2m and surface velocity of 20m/s, the average rate of deformation in natural debris flows is very small, i.e. $du/dz = 10/s$.

Progress has also been made in the analysis of both average and surface velocities. Hashimoto presents relationships for surface velocity u_s/u_* and mean velocity $\overline{u}/u_*$ proportional to h/d as shown in Eqs. 19, 20 and 27 where $\overline{u}$ is the mean velocity, u_s is the surface velocity, u_* is the shear velocity, h is the flow depth and d is the grain diameter. This analysis is quite intriguing, because given the grain size and shear velocity, not only the velocity profile but also the mean flow velocity increases linearly with flow depth. This linear model is in agreement with laboratory data at values of $h/d < 30$ has shown on Figures 7 and 13 of Hashimoto's paper. However, the experimental data deviates substantially from the inertial model of Hashimoto at values of $h/d > 30$. When $h/d > 30$, the reader may find better agreement with a turbulence equation of the type:

$\frac{\overline{u}}{u_*} = 575 \log \alpha \frac{h}{d}$, in which a value $\alpha = 1$ (compared to $\alpha = 12.2$ for clear water) fits the experimental observations of Hashimoto on both Figures 8 and 13, and to some extent in Figure 7, considering $u_s \cong 2\overline{u}$. In any event, the practitioner will notice that the mean debris flow velocity is less than that calculated with traditional turbulent flow equation ($\alpha = 12.2$).

One of the primary conclusions of this set of papers is that the inertial impact of particles cannot be dominant when $h/d > 30$. Hashimoto's conclusion also finds support in Takahashi's paper stating that the turbulent flow regime in natural sand and water mixtures appears when $h/d > 20$–30, with reference to Arai and Takahashi (1986). This important conclusion is very practical in that for debris flows where typical flow depths reach 2m, a particle size of at least 80mm is required to induce sufficient dispersive stress to overcome the turbulent stress. Consequently, natural debris flows of particle mixtures finer that 80mm (gravel, sand, silt and clay) remain either turbulent or viscous, but not dispersive.

In summary, significant progress has been made in recent years in understanding the dynamics of debris flows. Most shear stress components have been identified and several components can be estimated from available laboratory experiments. The quadratic rheological model seems effective; the quantitative evaluation of all components describing yield, viscous, turbulent and dispersive stresses is readily possible, although subject to refinement. Advances in the analysis of velocity profiles, surface and mean flow velocities lead to the conclusion that particle impact cannot be dominant when $h/d > 25$. The practitioner will find that velocity profiles are nearly linear and the rates of deformation are very small, of the order of $du/dz \cong 10/s$. The mean flow velocity is less than calculated with the standard turbulent flow equation with $\alpha = 12.2$.

Future improvements are possible through an accurate determination of: 1) viscosity as a function of the concentration of fine particles; 2) mixing length generating turbulent stress in hyperconcentrations; and 3) the coefficient f of the dispersive stress relationship. Experimental research on inertial impact of coarse gravel particles, naturally non-buoyant particles is in dire need. A better understanding of the effects of clay mineralogy and fine sediment concentration on the viscosity of a mixture will improve our understanding of the rheology of hyperconcentrations. Advances in our physical understanding of the dynamics of debris flows will enhance our ability to model hyperconcentrated flows, mud flows, and debris flows. These simulation models will in turn facilitate improved design of adequate countermeasures to protect living communities against devastating debris flows.

A Comparison Between Gravity Flows of Dry Sand and Sand-Water Mixtures

Haruyuki Hashimoto

Department of Civil Engineering
Kyushu University
Fukuoka 812, Japan.

Abstract

The flow model and constitutive equations proposed by Tsubaki, Hashimoto and Suetsugi (1982) are found valid at smaller values of the ratio of flow depth to sand grain size in the gravity flows of dry sand and hyperconcentrated mixtures of sand and water. Nondimensional parameters governing these flows are derived from the comparison between intergranular-stress and inertia terms in the momentum equations. Their flow behaviour can be explained by using the constitutive equations and the parameters.

1. Introduction

Debris flow is a gravity flow of hyperconcentrated mixtures of sediment and water. Such a flow generates stresses due to interaction among sediment grains and behaves like a non-newtonian fluid. The prediction of the flow requires the knowledge of the intergranular stresses, concentration and velocity. Therefore theoretical and experimental works on the constitutive equations and the flow behaviour of solid-liquid mixtures have been performed under active development (e.g. Bagnold 1954; Takahashi 1978; Tsubaki, Hashimoto and Suetsugi 1982; Ashida, Egashira, Kamiya and Sasaki 1985).

Bagnold (1954) performed a pioneering work on the intergranular stresses and presented the constitutive relationship similar to that in dilatant fluid. Takahashi(1978) applied Bagnold's model to the mixture flows of sand and water and derived the equations of velocity and concentration.

Tsubaki, Hashimoto and Suetsugi (1982) emphasized the importance of many-body collisions as the interaction among grains. By modelling the grain-grain interactions they derived equations for intergranular stresses and obtained the distribution of velocity and concentration. Ashida, Egashira, Kamiya and Sasaki (1985) introduced fluid stress as the role of liquid phase and yield stress as intergranular interactions.

On the other hand, there are gravity flows of dry granular materials similar to debris flow. Although many works on the granular flows have been done (e.g. Kanatani 1979; Ogawa, Umemura and Oshima 1980; Savage and Jeffrey 1981), most of them were focused on the constitutive relations and not on flow behaviour.

In the present work first we review the flow model of Tsubaki et al. (1982) as a typical model. Second we apply the model to the gravity flows of dry sand and sand-water mixtures. Finally we discuss the difference and similarity between these gravity flows and derive the nondimensional parameters governing flow situation.

2. The Flow Model of Tsubaki, Hashimoto and Suetugi (1982)

Tsubaki et al. (1982) made experiments to know the way of interaction among particles in shear flow of a solid-liquid mixture. They used nearly spherical particles of specific gravity $\sigma/\rho = 1.25$ and diameter $d = 17.4\,mm$ as a solid material and water as liquid. They produced the mixture flow at high concentration and high shear rate in an inclined flume. Close-up view of moving particles was taken with a $16\,mm$ high-speed camera running at 100 frames per second from the side of the flume. Analyzing the $16\,mm$ films, they examined the profiles of particle velocity and details of particle-particle interactions. This result is shown schematically in Fig. 1.

Particles approach a reference particle at a relative velocity which is related to the mean shear. We can choose the reference particle arbitrarily.

The particles collide against the 'upstream' quadrant on the reference particle. These particles then move over the ' upstream' quadrant and at last separate from the 'downstream' quadrant. Furthermore the colliding particles always have other particles in contact with them. The particles touching the colliding ones are also in contact with other particles. Tsubaki et al. called these collisions 'many- body' ones. They pointed out that 'many-body' collisions play a predominant role for momentum transfer in the shear flow of the hyperconcentrated mixture.

On the basis of the way of particle-particle interactions, they distinguished two different stresses; one is collision stress due to the collision force and the other is contact stress due to the force acting during contact after collision.

2.1 Collision Stress

They considered a collision between particles o and i of equal diameter, as shown in Fig. 2. The change in momentum of particle i is

$$m(\mathbf{u}_i' - \mathbf{u}_i) = (\mathbf{n} - \mu\,\mathbf{s}) \int_0^{t_c} F\,dt \tag{1}$$

$$= -m(\mathbf{u}_i \cdot \mathbf{n})(\mathbf{n} - \mu\,\mathbf{s}) \tag{2}$$

where m is the mass of each particle, $\mathbf{u}_i$ and $\mathbf{u}_i'$ are the relative velocity of particle i to particle o before and after the collision respectively, $\mathbf{n}$ and $\mathbf{s}$ are the unit vectors in the direction from the center of particle o to the collision point P and in the sliding direction of particle i respectively, μ is a coefficient of sliding friction, t_c is the collision time, and F is the normal collision force. In the derivation of Eq. (2) they used no-rebound condition $\mathbf{u}_i' \cdot \mathbf{n} = 0$ and orthogonal relation $\mathbf{n} \cdot \mathbf{s} = \mathbf{0}$.

For two-dimensional shear flow $u(z)$, $\mathbf{u}_i$ was written as

$$\mathbf{u}_i = (d \cos\theta \frac{du}{dz}, 0, 0) \tag{3}$$

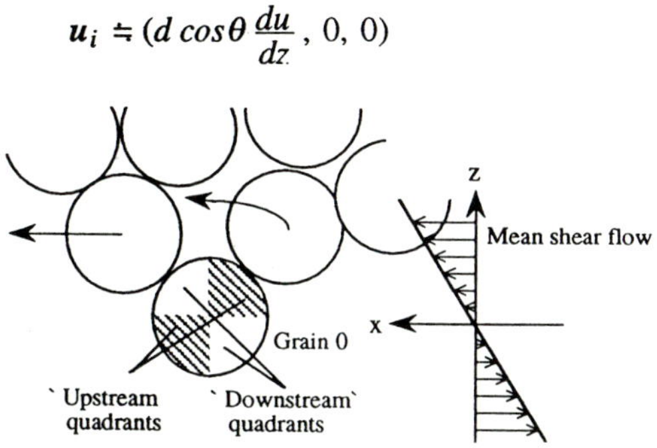

Fig. 1. Schematic diagram of the relative movement of grains

The number of collisions of particle o per unit time at angles within the ranges θ and $\theta + d\theta$, and ϕ and $\phi + d\phi$ is

$$dn = -\left(d^2 \sin\theta \; d\theta \; d\phi \; \mathbf{u}_i \cdot \mathbf{n}\right) N \tag{4}$$

where N is the number density of particles given by

$$N = \frac{C}{\frac{\pi}{6} d^3} = \beta \frac{C}{C_* \, d^3} \tag{5}$$

where $\beta = 1.15$.

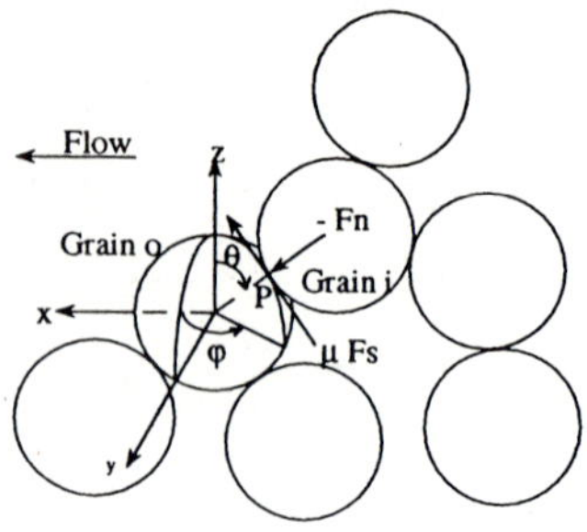

Fig. 2. Collision of grain o with grain i

Using the principle of action and reaction and integrating the rate of change of momentum, Tsubaki et al. expressed the mean collision force acting on particle o as

$$F_o = -\int m(\mathbf{u}_i' - \mathbf{u}_i)\ dn \tag{6}$$

The collision stress $\tau_\mathbf{z}$ acting on the plane S_z of unit area perpendicular to the z axis was derived as follows: (i) When a particle is cut at an angle θ' by the plane S_z as shown in Fig. 3, the collision force $\mathbf{F_z}(\theta)'$ acting on the shaded surface of the particle can be described by Eq. (6); (ii) the number of particles cut by S_z at angles within the ranges θ' and $\theta' + d\theta'$, can be given by $N(d/2)\ \sin\theta'\ d\theta'$.

Thus $\tau_\mathbf{z}$ is

$$\tau_z = \int_{\theta'=0}^{\pi} \mathbf{F_z}(\theta') N(d/2)\ \sin\theta'\ d\theta' \tag{7}$$

This stress is due to binary collisions and do not take account of the effect of many-body collisions. The momentum transfer in many-body collisions was evaluated as follows. Let us term particles such as particle i colliding with

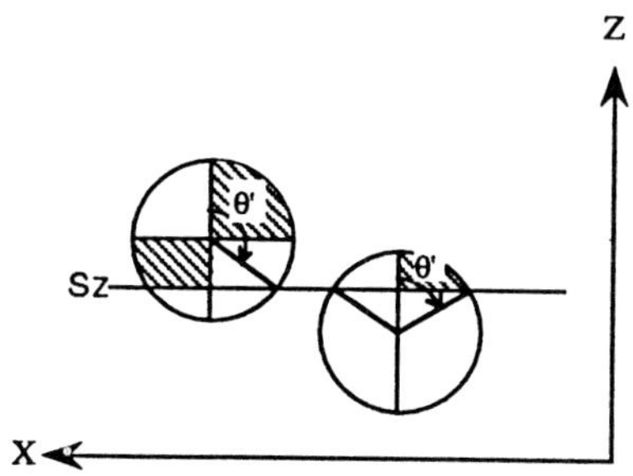

Fig. 3. Definition sketch of θ'

particle o '1st-order particle'. '2nd-order particle' is defined to be particles in contact with '1st-order one', '3rd-order particle' be particles in contact with '2nd-order one', and so on. If the collision between particles o and i happens, then the change in momentum of each particle is

$$m d\mathbf{u_1} = (\mathbf{n} - \mu\,\mathbf{s})F\,dt + \sum_{1N_2} \mathbf{E}_{2,1}\ dt$$

$$m d\mathbf{u_j} = \mathbf{E}_{j-1,j}\ dt + \sum_{jN_{j+1}} \mathbf{E}_{j+1,j}\ dt \qquad (j = 2, 3, \cdots, l) \tag{8}$$

where $d\mathbf{u}_j$ is the velocity change of jth-order particle, $\mathbf{E}_{j,j+1}$ is the force acting on $(j+1)$th- order particle from jth-order particle and jN_{j+1} is the number of $(j+1)$th-order particles in contact with jth-order one. Obviously, $\mathbf{E}_{j+1,j} = -\mathbf{E}_{j,j+1}$ and the following relations were assumed in Eq. (8).

$$F > \sum_{1N_2} |\mathbf{E}_{2,1}| \ , \quad and \quad |\mathbf{E}_{j,j+1}| > \sum_{j+1\,N_{j+2}} |\mathbf{E}_{j+2,j+1}|$$

$$\sum_{1N_2} d\mathbf{u}_2/d\mathbf{u}_1 = \sum_{2N_3} d\mathbf{u}_3/d\mathbf{u}_2 = \cdots = \sum_{jN_{j+1}} d\mathbf{u}_{j+1}/d\mathbf{u}_j = \varepsilon$$

where $0 < \varepsilon < 1$.

Summing the right and left sides of Eq. (8) over all the particles, using the above relations, and putting $l \to \infty$, Tsubaki et al. obtained

$$\frac{m}{1-\varepsilon}(\mathbf{u}_i' - \mathbf{u}_i) = (\mathbf{n} - \mu\mathbf{s}) \int_0^{t_c} F\,dt \tag{9}$$

Comparing Eq. (1) with Eq. (9), they found that the mass of a particle in hyperconcentrated mixture flow must be replaced by $M = m/(1-\varepsilon)$. Since ε approaches 1 as $C \to C_*$, ε was expanded in a Taylor series about $C = C_*$.

$$\varepsilon = 1 + \frac{1}{k_M}\frac{C - C_*}{C_*} \tag{10}$$

where k_M is an experimental coefficient. To a first approximation, M was expressed as

$$M = k_M\ m/(1 - C/C_*) \tag{11}$$

2.2 Contact Stress

Contact stress was determined by both the number of contact points on a single particle and contact force exerted on it from a neighbouring particle. Referring to Eq. (7) and Fig. 3 and neglecting contact shear stress because of its minor role, Tsubaki et al. expressed contact pressure p as $p = (C_* n_c' F')C/C_*$. Here n_c' is the density of contact points on the particle, and F' is the normal contact force. They assumed $n_c'F'$ to be produced by the excess immersed weight of particles which the normal collision stress cannot support and be related to the concentration. In gravity flow with a free surface, the pressure must vanish as $C \to C_S$, where C_S is the concentration at the surface. $C_* n_c' F'$ was expanded in power series of C about $C = C_S$. The following relations were deduced.

$$p = K_p \frac{C}{C_*}\frac{C - C_S}{C_S} \quad , \quad \frac{K_p}{C_S(\sigma - \rho)\, g\, h\, \cos\theta_0} = \chi \tag{12}$$

where h is the flow depth and χ is an experimental coefficient. Thus the stress components became

$$\begin{aligned} \sigma_{jk} &= \tau_{jk} - p\delta_{jk} \\ &= \beta^2\, k_M \frac{m}{d}\frac{(C/C_*)^2}{1 - C/C_*}(\frac{du}{dz})^2\, A_{jk} - K_p \frac{C}{C_*}\frac{C - C_S}{C_S}\delta_{jk} \end{aligned} \tag{13}$$

where A_{jk} is a tensor expressed by friction coefficient μ, and δ_{jk} is the Kronecker delta.

3. Gravity Flows of Dry Sand

Savage (1979) performed experiments of gravity flows of polystyrene beads in an inclined chute. He found that the velocity profiles have an inflection point, although the chute was very small, that is, 3.86 cm wide and 1.22 m long. The relative flow depth was $h/d = 10 \sim 12$. Ishida, Hatano and Shirai (1980) made experiments of chute flows in which the particles were fluidized by air flow through a porous bed. Savage (1984) reviewed the papers on various types of flow of granular materials. Kitou, Hirano and Hashimoto (1993) made experiments of flows of dry sand in an inclined open channel. They used three kinds of channel bed; one was a fixed bed roughened with same sand as flowing sand and the others were smooth fixed beds of plywood and acrylic board. The size of the roughness of plywood and acrylic board was nearly same as that of fine sand and silt, respectively. From these works it is found that flow situation changes with bed slope and roughness.

First, we can distinguish three types of flow termed quasi-static, laminar and dispersive types on the basis of bed slopes.

1. Quasi-static type of flow denotes slow flows which occur at bed slopes close to the angle of repose of sand. The determination of flow depth is difficult because of sand deposition on top of the bed. Flow situation is unsteady. This has small shear rates and high concentration of sand.
2. Laminar type of flow denotes rapid flows which occur at the steeper slopes. Since no deposition occurs and free surface is clear, the determination of flow depth is easy. In the case of coarser materials the flow behaves like a laminar flow of fluid.
3. Dispersive type of flow denotes rapid flows which occur at further steeper slopes. Saltation of sand grains is vigorous. Since grain concentration becomes lower near free surface and the surface is not clear, the determination of flow depth is not easy.

Second, we can distinguish shear and uniform flow corresponding to bed roughness. In the case of smooth bed flowing grains slip at the bottom and their velocity profile becomes uniform, while in the case of rough bed larger shear occurs because of no slip at the bottom.

In the present chapter we focus the discussion on the laminar type in the case of fixed bed of roughness of same size as the size of flowing material; this corresponds to the boundary condition of $u = 0$, $C = C_*$ at $z = 0$.

Consider a two-dimensional steady and uniform flow, as shown in Fig. 4. The momentum equation can be written as

$$\sigma C g \sin\theta_0 + \frac{\partial \sigma_{zx}}{\partial z} = 0 \tag{14}$$

$$-\sigma C g \cos\theta_0 + \frac{\partial \sigma_{zz}}{\partial z} = 0 \tag{15}$$

where air density ρ is neglected because of its minor role compared with grain density σ.

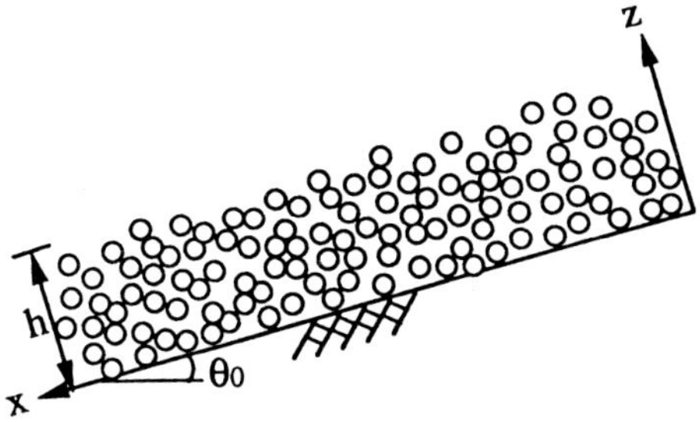

Fig. 4. Schematic diagram of gravity flow of dry sand

Substituting Eq. (13) into Eqs. (14) and (15) and using the boundary condition $\eta = z/h = 0$, $C = C_*$, we can derive concentration profile

$$\eta = \left[2\left(1 - \frac{C}{C_*}\right) + \frac{C_s}{C_*} \ln \frac{C}{C_*} \right] \frac{\chi}{\psi} \tag{16}$$

where ψ and χ are defined by

$$\psi = \frac{\gamma - \tan\theta_0}{\gamma} \; , \; \chi = \frac{K_p}{C_S \, \sigma \, g \, h \, \cos\theta_0} \; , \; \gamma = -\frac{\tau_{zx}}{\tau_{zz}} = \frac{0.0762 - 0.102\mu}{0.0898 - 0.067\mu} \tag{17}$$

From Eq.(14) we can obtain velocity profile under the condition of $u = 0$ at $z = 0$ as

$$\frac{u}{u_S} = \frac{\displaystyle\int_C^{C_*} F_g(C)\, dC}{\displaystyle\int_{C_S}^{C_*} F_g(C)\, dC} \tag{18}$$

where surface u_s symbol can be written as

$$\left.\begin{aligned} \frac{u_s}{u_*} &= \frac{1}{\sqrt{K_M}}\frac{h}{d}\left(\frac{\chi}{\psi}\right)^{3/2}\frac{1}{C_*}\int_{C_S}^{C_*} F_g(C)\,dC \\ K_M &= \tfrac{\pi}{6}(0.0762+0.102\mu)\beta^2\,k_M \\ F_g(C) &= \frac{2C-C_S}{C}\sqrt{(C-C_S)\frac{1-C/C_*}{C/C_*}} \end{aligned}\right\} \tag{19}$$

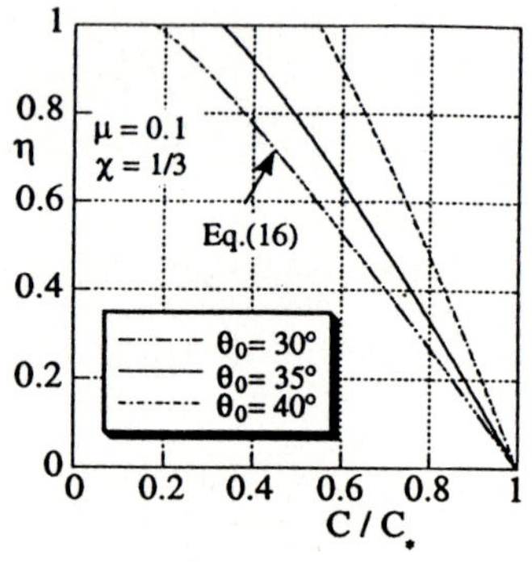

Fig. 5. Concentration profile of dry sand

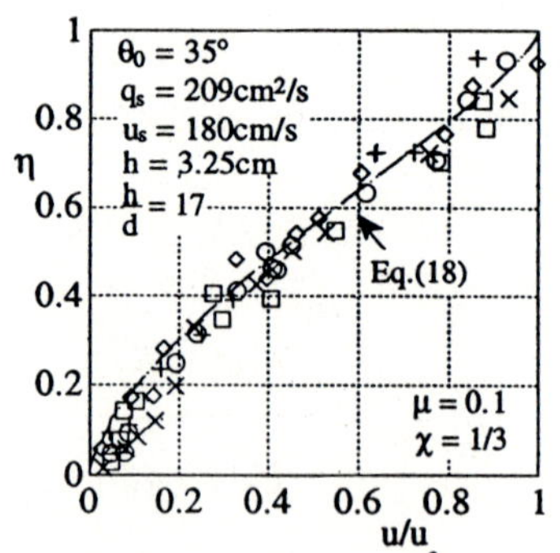

Fig. 6. Nondimensional velocity profile of dry sand

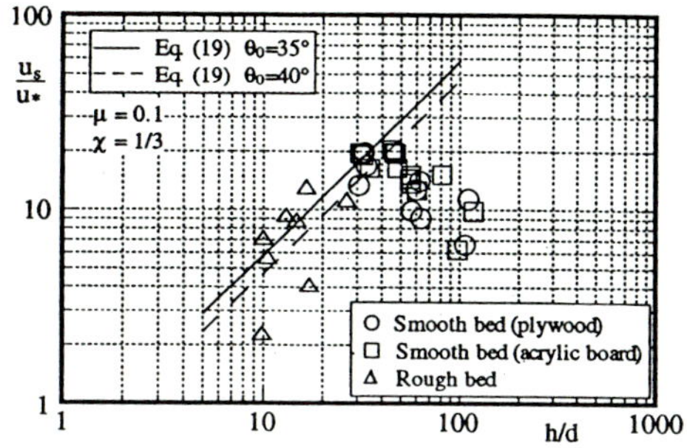

Fig. 7. Variation of nondimensional surface velocity with relative flow depth

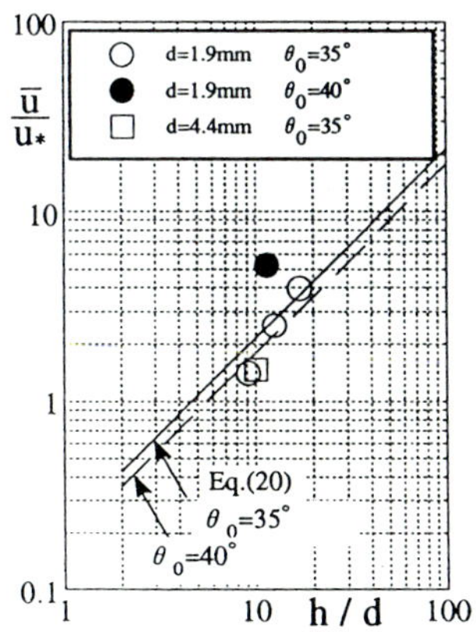

Fig. 8. Variation of nondimensional average velocity with relative flow depth

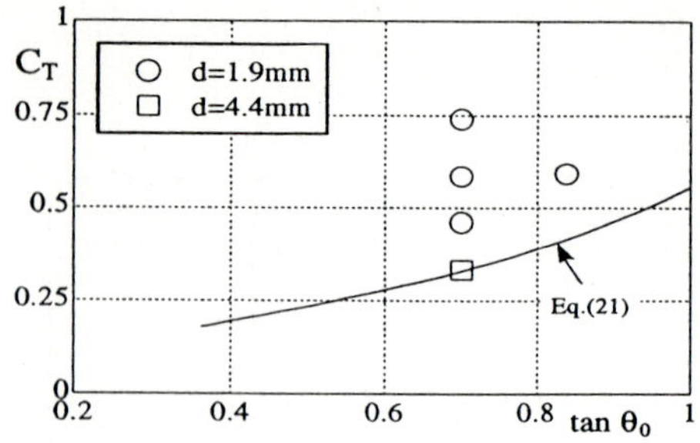

Fig. 9. Flux-averaged concentration versus bed slope

Further integrating Eq. (18) yields

$$\frac{\overline{u}}{u_*} = \frac{1}{\sqrt{K_M}}\frac{h}{d}\left(\frac{\chi}{\psi}\right)^{5/2}\frac{1}{C_*^2}\int_{C_S}^{C_*}\frac{2C - C_S}{C}\left(\int_{C}^{C_*} F_g(C)\,dC\right)dC \tag{20}$$

Flux-averaged concentration becomes

$$C_T = \frac{\displaystyle\int_0^1 C\,U\,d\eta}{\displaystyle\int_0^1 u\,d\eta} \tag{21}$$

The calculation of concentration profile is shown in Fig. 5. Here we use the values of μ and χ determined for the mixture flow by Tsubaki et al. (1980. The comparison between the calculations of Eqs. (18), (19), (20) and (21) and the experiments (Kitou et al. 1993) is made in Figs. 6, 7, 8 and 9. In the region of smaller values of h/d they agree, while in the region of larger values of h/d the do not agree (see Fig. 7). This problem is discussed in Chapter 6.

4. Gravity Flows of Hyperconcentrated Mixtures of Sand and Water

Tsubaki et al. (1982) made experiments of the gravity flows of solid-water mixtures in a sloping flume by using coarse sand of $\sigma/\rho = 2.59$ and $d = 4.6\,mm$ and artificial coarse particles of σ/ρ= 1.6 and d= 8.2 mm as solid materials. They measured velocity profile, flux- averaged concentration, average velocity and flow depth under the condition of movable bed. Furthermore, Hirano, Hashimoto et al. (1992) performed experiments with sand of various sizes under three kinds of bed condition, that is, movable bed and fixed beds of plywood and acrylic board. The roughness of plywood and acrylic board corresponds to the size of fine sand and silt, respectively. They measured average velocity and velocity profile for the mixture flows. They discussed the effect of bed condition on velocity profile and flow resistance. In the present chapter we focus the discussion on the flows over movable bed and fixed bed of roughness of same size as the size of flowing material. In this case we can use the boundary condition of $u = 0$, $C = C_*$ at $z = 0$.

Consider a two-dimensional steady and uniform flow of the mixtures, as shown in Fig. 10. Since the flow contains interstitial water and high concentration of sand grains, it must be discussed as two-phase flow. Therefore we can write the momentum equations as

$$\{\sigma C + \rho(1 - C)\}\, g \sin\theta_0 + \frac{\partial \sigma_{zx}}{\partial z} = 0 \tag{22}$$

$$-(\sigma - \rho)\, C\, g\, \cos\theta_0 + \frac{\partial \sigma_{zz}}{\partial z} = 0 \tag{23}$$

where Eq. (22) is for both the phases and Eq. (23) for grain phase. In Eq. (22) Reynolds stress due to the turbulence of the interstitial water is eliminated because of its minor role compared with the intergranular stress.

For convenience introduce the coefficients defined by

$$\left.\begin{aligned} &\frac{\tau_{zx}}{\tau_{zz}} = -\frac{\gamma}{1-\rho(2\sigma)} \equiv -\alpha \ , \ C_\alpha = \frac{\tan\theta_0}{\frac{\sigma - rho}{\rho}(\alpha - \tan\theta_0}) \ , \\ &\psi = \frac{\alpha - \tan\theta_0}{\alpha} \ , \ \chi = \frac{K_p}{C_S\,(\sigma-\rho)\,g\,h\cos\theta_0} \ , \end{aligned}\right\} \qquad (24)$$

Substituting Eq. (13) into Eqs. (22) and (23) and combining Eqs. (22) and (23) gives the distributions of grain concentration and velocity. The distribution of concentration becomes

$$\eta = \left[2\left(1-\frac{C}{C_*}\right) - \frac{2C_\alpha - C_S}{C_*}\ln\frac{C-C_\alpha}{C_*-C_\alpha}\right]\frac{\chi}{\psi} \qquad (25)$$

where $\eta = z/h$ and C_S can be evaluated by putting $C = C_S$ at $\eta = 1$. Using shear velocity $u_* = (g\,h\,\sin\theta_0)^{1/2}$ and surface velocity u_s, we can write the nondimensional form of velocity profile as

$$\frac{u}{u_S} = \frac{\int_C^{C_*} F_d(C)\,dC}{\int_{C_S}^{C_*} F_d(C)\,dC} \qquad (26)$$

where $F_d(C)$ is a function of concentration C.

Average velocity $\overline{u}$ and flux-averaged concentration C_T become

$$\frac{\overline{u}}{u_*} = \frac{h}{d}\left(\frac{\sigma}{\rho}K_M\left(\frac{\psi}{\chi}C_*\right)^5\right)^{-1/2}\int_{C_S}^{C_*}\frac{2C-C_S}{C-C_\alpha}\left(\int_C^{C_*}F_d(C)\,dC\right)dC \qquad (27)$$

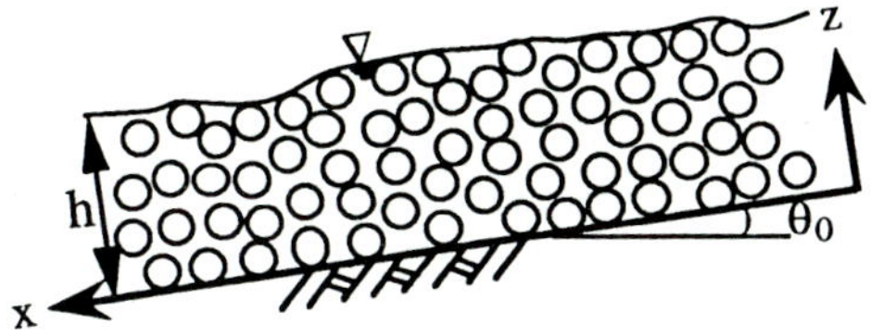

Fig. 10. Schematic diagram of gravity flow of a sand-water mixture

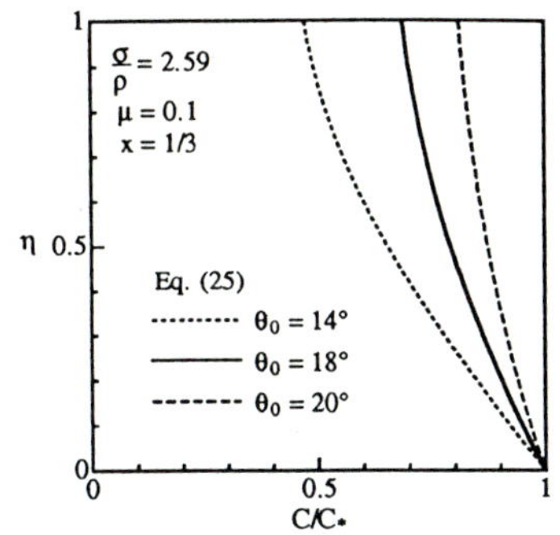

Fig. 11. Concentration profile for the mixture flow

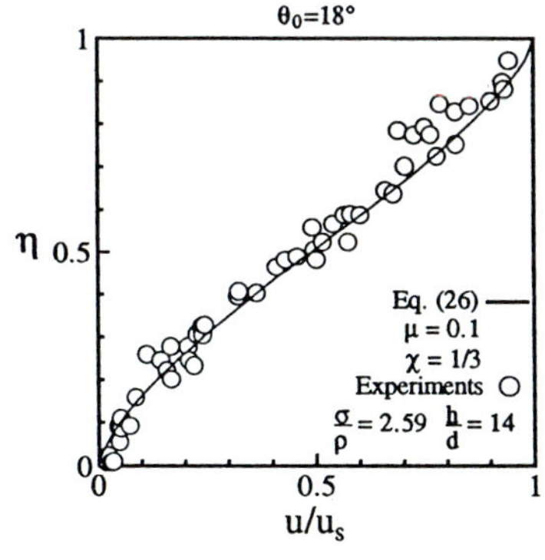

Fig. 12. Nondimensional velocity profile for the mixture flow

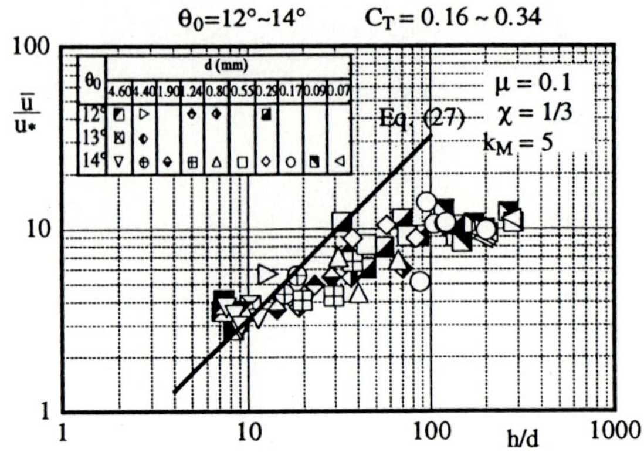

Fig. 13. Variation of nondimensional average velocity with relative flow depth

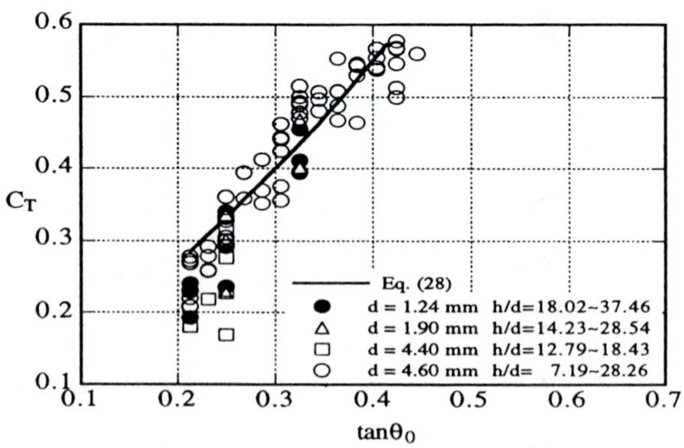

Fig. 14. Flux-averaged concentration versus bed slope

$$C_T = \frac{\int_0^1 C\,U\,d\eta}{\int_0^1 u\,d\eta} \tag{28}$$

Concentration profile calculated by Eq. (25) is shown in Fig. 11. Here the values of μ and χ are same as those for dry-sand flow. Figs. 12, 13 and 14 show the comparison of the calculations of Eqs. (26), (27) and (28) with the experiments. They agree in the region of smaller values of h/d. In the region of the larger values of h/d, however, there is a gap between the calculations and the experimental results (see Fig. 13). This problem is discussed in Chapter 6.

5. A Comparison Between the Gravity Flows of Dry Sand and Sand-Water Mixtures

The difference between the gravity flows of dry sand and sand-water mixtures is due to fluid density ρ. The basic equations for the former flows neglect fluid density while the equations for the latter consider the fluid density. As a result C_α defined by Eq. (24) appears in the equations for the mixture flow and doesn't for the dry-sand flows; C_α is a function of θ_0 and $(\sigma - \rho)/\rho$ which increases with θ_0. Comparing Fig. 5 with Fig. 11, we can see the effect of C_α on the concentration profile.

On the other hand, similar characteristics of velocity profile, surface velocity and average velocity are found between dry sand and the mixture. In particular, surface and average velocity are proportional to h/d for the both flows within the range such that $h/d < 20 \sim 30$.

6. Parameters Governing Flow Situation

Reynolds number expresses the ratio of inertial to viscous forces and is useful in the description of situation of clear water flow. Corresponding to Reynolds number there must be nondimensional parameters for the hyperconcentrated flows, such as the mixture flows and dry-sand flows. Hashimoto and Hirano (1992) discussed this problem by comparing friction forces on a bed and drag forces on a tube from the mixture flows. Since the equations of Tsubaki et al. are found valid for the flows of dry sand and the mixtures, we can discuss the universal parameter for the these flows. The momentum equation in the x-direction for two-dimensional steady flows is

$$u\frac{\partial u}{\partial x} + \nu\frac{\partial u}{\partial z} = -\frac{1}{\rho_t}\frac{\partial p}{\partial x} + \frac{1}{\rho_t}(\frac{\partial}{\partial x}\frac{\partial \tau_{zx}}{\partial z} + \frac{\partial \tau_{xx}}{\partial x}) \tag{29}$$

where u and v are velocity components in the x and z directions, p is pressure including gravity components of the flows and contact stress, τ_{zx} and τ_{xx} are collision stresses and $\rho_t = \sigma_C + \rho(1_C)$ is density of flows. Here Reynolds stress due to the turbulence of the interstitial fluid is assumed negligibly minor compared to the collision stresses.

U being a characteristic velocity and L being a characteristic length, we can estimate the inertia terms as U^2/L and the collision-stress terms as $d^2F(C)U^2/L^3$. Here $F(C)$ is a function which increases with C. The ratio of these terms is

$$\frac{U^2/L}{d^2\ F(C)\ U^2/L^3} = \frac{1}{F(C)}(\frac{L}{d})^2 \tag{30}$$

Thus L/d and C are found important parameters. At larger values of C and smaller values of L/d intergranular-stress terms play major role compared with the inertia terms. At smaller values of C and larger values of L/d, on the other hand, the inertia terms become important relatively to the intergranular-stress terms.

Therefore the gap between the calculations and the experiments in the region of larger values of h/d in Figs. 7 and 13 can be explained as follows: at the larger values inertial forces become dominant compared with the intergranular force. This corresponds to the work of Arai and Takahashi (1986), although they did not explain the physical meaning of relative flow depth h/d.

7. Conclusions

It is found that the flow model and the constitutive equations of Tsubaki et al. can explain the flow behaviour of dry sand and hyperconcentrated mixtures of sand and water within the range such that $h/d < 20 \sim 30$. In this range intergranular forces play major role. In the range such that $h/d > 100$, on the other hand, inertial forces become dominant compared with the intergranular forces. In the intermediate range, that is $20 \sim 30 < h/d < 100$, the effect of the both forces is important.

REFERENCES

Arai, M. and Takahashi, T. (1986). " The Mechanics of Mud Flow ", Pro. JSCE, No. 375/II-6.

Ashida, K., Egashira, S., Kamiya, H. and Sasaki, H. (1985). " The Friction Law and Moving Velocity of Soil Block on Slope, " Ann. Disaster Prev. Res. Inst. Kyoto Univ., No. 28 B-2.

Bagnold, R.A. (1954). " Experiments on a Gravity-Free Dispersion of Large Solid Sphere in a Newtonian Fluid under Shear ", Proc. Roy. Soc. A, Vol. 225.

Hashimoto, H. and Hirano, M. (1992). " Rapid Flows of Sand-Water Mixtures at High Concentration in a Steep Channel ", Advances in Micromechanics of Granular Materials, H.H. Shen et al. (Editors), Elsevier Science Publications B.V.

Hirano, M., Hashimoto, H., Fukutomi, A., Taguma, K. and Pallu, M.S. (1992). " Nondimensional Parameters Governing Hyperconcentrated Flow in an Open Channel ", Proc. Hydraulic Engineering, JSCE, Vol. 36.V

Ishida, M., Hatano, H. and Shirai, T. (1980). " The Flow of Solid Particles in an Aerated Inclined Channel ", Powder Technol. Vol. 27.

Kanatani, K. (1972). " A Micropolar Continuum Theory for the Flow of Granular Materials ", Int. J. Eng. Sci., Vol. 17.

Kitou, K., Hirano, M. and Hashimoto, H. (1993). " Characteristics of Granular Flow in an Inclined Open Channel ", Proc. Hydraulic Engineering, JSCE, Vol. 37.
Ogawa, S., Umemura, A. and Oshima, N. (1980). " On the Equations of Fully Fluidized Granular Materials ", Z. argew. Math. Phys., Vol. 31.
Savage, S. B. (1979). " Gravity Flow of Cohesionless Granular Materials in Chutes and Channels ", J. Fluid Mech., Vol. 92.
Savage, S. B. (1984). " The Mechanics of Rapid Granular Flows ", Advances in Applied Mechanics, Vol. 24.
Savage, S. B. and Jeffrey, D. J. (1981). " The Stress Tensor in a Granular Flow at High Shear Rates ", J. Fluid Mech., Vol. 110.
Takahashi, T. (1978). " Mechanical Characteristics of Debris Flow ", J.H.D, ASCE, Vol. 104, HY8.
Tsubaki, T., Hashimoto, H. and Suetsugi, T. (1982). " Grain Stresses and Flow Properties of Debris Flow ", Proc. JSCE, No.317.

DISCUSSION

Michiue: Is the diferent point between the dry sand flow and the mixture flow of sand and water to neglect only the buoyancy force in the dry sand flow to obtain the velocity distribution of it? Are the other terms the same condition of both flows?

Hashimoto: If we use our constitutive equation, neglecting air density in the dry-sand flow and considering water density in the mixture flow yields the equations of concentration and velocity for each flow. But the other conditions are same.

Takahashi: 1. The shape of solids concentration does not affect much for the velocity profile, doesn't it?

2. Is it necessary to change the constitutive equation for lower dense region and for the upper thin concentration region?

Hashimoto: 1. A decrease in solid concentration makes the velocity profile have inflection point.

2. It depends on the condition. Under the condition of $C =\geq 0.3$ and high shear, our equation is enough.

Egashira: 1. p, contact force, disappears when the sediment distributes uniformly vertically, according to Eq. (12). Is it right?

2. Why energy dissipation does not occur due to the deformation of contact field? Is it natural that the contact stress contributes to the shear stress?

Hashimoto: 1. p disappears in the case of uniform distribution of concentration.

2. Our idea shows many-body collisions are dominant for energy dissipation under the condition of high shear and high concentration. In this case shear stress due to many-body collisions are mainly produced.

Julien: In Eq.(27), $\mathbf{uu}_* \propto hd$ (1)

on Fig. 13, the laboratory measurements indicate

$\mathbf{uu}_* \propto (hd)^{0.5}$ (2)

1. Can you explain the discrepancies between Eqs.(1) and (2)?

2. Is there any laboratory data in the range $h/d > 300$ available in the literature. Can you plot them on Fig.13?

Hashimoto: 1. There is no discrepancies between Eq.(27) and the laboratory measurements. As discussed in Chapter 6, intergranular forces play major role within the range such that $h/d < 20\ 30$. In the range such that $h/d > 100$, on the other hand, inertial forces become dominant compared with the intergranular forces. In the intermediate range, that is $20\ 30 < h/d < 100$, the effect of the both forces is important. Eq.(2) expresses approximately average velocity in the range such that $20\ 30 < h/d < 100$. But Eq.(1) is for the range such that $h/d < 20\ 30$.

2. There is no laboratory data in the range $h/d > 300$. We

discuss the flows at high concentration. It is very difficult to produce such flows in the range such that $h/d > 300$ in the experiments.

Review Dynamic Modeling of Debris Flows

Chyan-Deng Jan[1] and Hsieh Wen Shen[2]

[1] Department of Hydraulics and Ocean Engineering
National Cheng Kung University
Tainan, Taiwan 70101, R.O.C.

[2] Department of Civil Engineering
University of California
Berkeley, CA 94720, USA

Abstract

A debris flow is a flow of sediment-fluid mixture. Four key features in momentum exchange of debris flows are fluid viscosity, turbulence, particle sliding friction, and particle collision. Debris flows were qualitatively classified into six flow regimes, according to the dominance of these key features. Existing rheological models for debris flows in various flow regimes were briefly reviewed. The characteristics of flow velocities for a steady, two-dimensional uniform debris flow in each flow regime were obtained by treating the debris flow as a single-layer uniform mixture. The mixed-layer models and the hydraulics of debris flow were also discussed.

1. Introduction

Debris flow is generally described as the gravity flow of soil, rocks, water and/or air mixture initiated by landslides with high runoff water flow. Its flow properties vary with water and clay content, sediment size and size distribution. The occurrence of debris flow is rather unpredictable and very destructive. Debris flows could move faster than the more common landslides and tend to affect areas at much greater distance from the source of hazard. Debris-flow disaster has been recognized as a critical problem facing the world today, and hence this has resulted in a dramatic increase in the number of studies of debris flow initiation and its flow phenomena. Debris flow is usually treated as the movement of a continuum for simplicity, in spite of the existence of solid particles in it. Since the mixture of debris is treated as a continuum, the equations of mass and momentum conservation for debris flow are similar to those for general fluid flow.

$$\frac{D\rho_m}{Dt} + \rho_m \frac{\partial u_k}{\partial x_k} = 0 \tag{1}$$

$$\rho_m \frac{\partial u_j}{\partial t} + \rho_m \, u_k \frac{\partial u_j}{\partial x_k} = \rho_m \, f_j + \frac{\partial \sigma_{ij}}{\partial x_j} \tag{2}$$

where ρ_m is the density of debris mixture and $\rho_m = (1 - C)\rho_f + c\,\rho_s$ for saturated debris mixture, in which ρ_f and ρ_s are densities of interstitial fluid and the solid particles in the debris mixture, respectively; t is time; x_k is the Cartesian coordinates and $k = 1, 2,$ and 3 , representing x, y and z directions; u_k is velocity components; f_j is the body-force components; σ_{ij} is the stress tensor in which the first subscript indicates that the stress component acts on the plane x_i=costant and the second subscript indicates that it acts in the x_j direction. The stress tensor σ_{ij} is usually expressed as $\sigma_{ij} = -p\,\delta_{ij} + \tau_{ij}$, where p the thermodynamic pressure; τ_{ij} is the shear-stress tensor and δ_{ij} is the Kronecker delta. Solid particles in debris flow can collide, rub, rotate, and vibrate as they translate downslope. Four key features in momentum exchange of debris flows are fluid viscosity, turbulence, particle sliding friction, and particle collision (Jan, 1992). Therefore, debris flows may exhibit non-Newtonian behavior, and thus rheological models (or constitutive equations) relating stress, strain, time and other variables are needed for debris-flow routing. In the last few decades, attempts to understand the physical processes in debris flow have received considerable attention and various rheological models have been experimentally and theoretically proposed (Bagnold, 1954; Savage, 1984; Shen, 1982). However, most of the models are limited in a two-dimensional debris flow, and each model has its own limit in application. For the sake of simplicity of discussion on the applicability of various models, debris flows are qualitatively classified into six flow regimes in this paper according to the dominance of these key features. The characteristics of flow velocities for a steady, two-dimensional uniform debris flow in various flow regimes can be obtained according to these models. A definition sketch for a steady 2-D uniform debris flow is shown in Fig. 1.

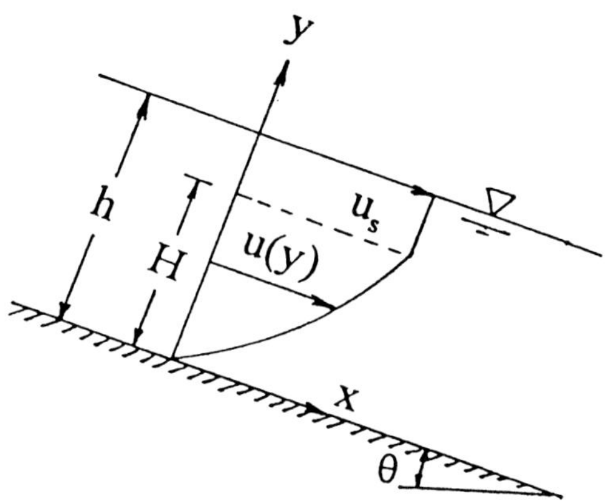

Fig. 1. Definition of sketch for a 2-D uniform debris flow

2. Rheological Models

2.1 Debris Flow in Friction Regime

Generally, the momentum exchange in a very slow movement of granular materials arises primarily from the mutual contact between particles, and any momentum exchange due to the interstitial fluid, is negligible. In such a flow, when the particles move, they closely stick together and deformation of the assembly of particles is slow. Particles sustain contact as they slide relative to one another for long times and momentum exchange is mainly caused by persistent rubbing of particles against their neighbors. The stresses in the flow are not governed by the magnitude of the applied rate of deformation, but mainly by the boundary stresses. A flow regime in this case is called the friction regime, plastic regime, or quasi-static regime (Savage, 1984; Johnson *et al.*, 1990). Flows in this regime exhibit the properties of a plastic material and lack a one-to-one correspondence between stresses and strain rate. In the last few decades many investigators have applied some theories of metal plasticity to the development of a theory for granular materials. An essential ingredient of these theories is a strongly pressure-dependence yield condition, such as the Coulomb theory which states that at a point of a granular solid plastic flow occurs when on any section the shear stress τ and the normal stress σ satisfy the relationship:

$$\tau = \tau_c + \sigma \tan\phi \tag{3}$$

in which τ_c and ϕ are the cohesion and the angle of internal friction of the bulk granular materials, respectively. It is usual to take ϕ as constant for mass of granular although it is well known to be dependent on the strain and thus on the solid concentration. However, the yield condition (Eq. 3) does not directly give any information about the kinematics of motion of granular material at yield. Actually, this is provided by the flow rules or velocity equations (Prakash and Rao, 1991; and Spencer, 1964/82). The friction model has been used with some kind of success to predict the stresses and velocity distributions in bins and hoppers as well as in the slow movement of soil. However, the applicability of the friction model to debris flow or granular flow with higher deformation rate is uncertain (Savage, 1979).

2.2 Debris Flow in Collision Regime

In contrast to the friction regime, when solid particles are widely spaced and deformation is rapid, contacts are of short duration and the momentum is mainly transferred by particle collisions. A flow regime in this case is called the collision regime, fully dynamic regime or grain-inertia regime (Bagnold, 1954; Savage, 1984; Johnson et al., 1990). The momentum transferred by particle collision has been found to be proportional to the square of the shear rate (Bagnold, 1954; Shen & Ackermann, 1982; Jenkins & Savage, 1983; Haff, 1983).

$$\tau = \alpha \left(\frac{du}{dy} \right)^2 \tag{4}$$

where α is a coefficient strongly depending on the density, size and size distribution, and concentration of solid particles as well as the internal friction angle of the granular mixture. Eq. 4 is originally proposed by Bagnold (1954) and usually called as dilatant model or dispersive model. From rotating-drum experiments with neutrally buoyant 1.32 mm spheres, Bagnold found that

$$\alpha = a_1 \, \rho_s \lambda^2 \, d^2 \, \sin \phi_d \tag{5}$$

where a_1 =empirical constant, ρ_s =particle density, d =particle diameter, ϕ_d =dynamic angle of internal friction, and λ is a linear grain concentration defined by Bagnold (1954) as the ratio of the grain diameter to the mean free dispersion distance and is related to the sediment volume concentration C and the maximum sediment concentration C_m.

$$\lambda = \left[(C_m/C)^{1/3} - 1 \right]^{-1} \tag{6}$$

in which C_m is the maximum value of C when all the grains are in static contact ($C_m = 0.74$ for the closest possible packing of uniform spheres). Bagnold also evaluated the empirical constant a_1 as 0.042. However, Takahashi (1980) directly applied Bagnold's equation in analysis of a steady debris flow down an inclined flume in laboratory and found a_1 equivalent to 0.5. This order-of-magnitude difference may indicate that (1) other factors, such as turbulent shear stress which is also proportional to the shear rate squared, may not be adequately incorporated in the above equation, and (2) Bagnold's results obtained from gravity-free flows may not directly apply to gravity flow (such as flow down an incline).

Since Bagnold's experimental work, many investigators, such as Savage and McKeown (1983), Savage and Sayed (1984), and Hanes and Inman (1985), have conducted similar experiments. Their results agree with that obtained by Bagnold in quality but not in quantity. Despite that some empiricism was involved in Bagnold's theoretical treatments, his model has been used as a theoretical basis in the development of constitutive relationships for rapid granular flows by Shen (1982), Shen and Ackermann (1982), Pasquarell *et al.* (1988), among others. Theoretical results can determine explicitly the coefficient α , but there is an order-of-magnitude difference between the theoretical and experimental results (Shen & Ackermann, 1982). If α is constant, the flow velocities for a steady, uniform two- dimensional flow down an inclined plane were obtained using Eq. 4. Table I shows the characteristics of velocity and velocity distribution for a 2-D uniform debris flow in the collisional regime. In Table I, θ is the inclined angle of the plane; h is the flow depth; u_s is the velocity at free surface; U is the depth-averaged velocity. The average velocity equals three fifths of the surface velocity. Based on the velocity distribution, the momentum correction coefficient β was found to be 1.25 for flow in this regime.

Velocity distribution	$u = \frac{2}{3}\sqrt{\frac{\rho_m\, g\, \sin\theta}{\alpha}}\,[h^{1.5} - (h-y)^{1.5}]$
Surface velocity	$u_s = \frac{2}{3}\sqrt{\frac{\rho_m\, g\, \sin\theta}{\alpha}}\,h^{1.5}$
Velocity defect	$\frac{u_s - u}{u_s} = (1 - \frac{y}{h})^{1.5}$
Average velocity	$U = \frac{2}{5}\sqrt{\frac{\rho_m\, g\, \sin\theta}{\alpha}}\,h^{1.5} = \frac{3}{5}u_s$
Momentum correction factor	$\beta = 1.25$

Table I Velocities for 2-D uniform debris flow in collisional regime

2.3 Debris Flow in Friction-Collisional Regime

As mentioned in the previous sections, stresses within a granular material are transmitted by forces exerted at points of mutual contacts between particles. When the assembly of particles is widely spaced in flow, individual contacts of particles are of short duration and particle collisions are dominant in momentum exchange. On the other hand, for slow deformation at high solid concentration, contacts are semi- permanent and sliding contacts between particles play a significant role in momentum exchange. The constitutive relations are available for these two limiting situations as discussed in last two sections. However, most situations of practical interest fall in the range between these extremes where both collisions and sliding friction are significant. Some constitutive equations for flowing granular materials in the intermediate regime have been proposed by investigators (such as McTigue,

1982, Johnson & Jackson, 1987) that consist of a friction part and a collision part. A constitutive equation proposed by McTigue (1982) for flow of a granular material down a slope is taken as an example, that is

$$\tau = \tau_c \cos\phi + \eta_1 (C^2 - C_o^2) \sin\phi + \eta_2 (C_m^2 - C^2) \left(\frac{du}{dy}\right)^2 \tag{7}$$

where η_1 and η_2 are coefficients to be determined; C_o and C_m are the minimum and the maximum solid volume concentrations, respectively. The sum of the first two terms on the right-hand side of Eq. 7 represents the yield stress τ_y which should be overcome before flow occurs. The stress-strain rate relation for flow in the friction-collisional regime can be generally expressed as

$$\tau = \tau_y + \alpha \left(\frac{du}{dy}\right)^2 \tag{8}$$

If τ_y and α are constant, the corresponding velocities of a 2-D steady uniform debris flow are summarized in Table II, in which H is the height from the bottom of the flow to the point where the applied shear stress equals the yield stress, and $h - H = [\tau_y/(\rho_m\, g \sin\phi)]$ is the thickness of plug in which no relative velocity exists.

Velocity distribution	$u = \frac{2}{3}\sqrt{\frac{\rho_m\, g\, \sin\theta}{\alpha}}[H^{1.5} - (H-y)^{1.5}]\ for\ 0 \le y \le H$
Surface velocity	$u_s = \frac{2}{3}\sqrt{\frac{\rho_m\, g\, \sin\theta}{\alpha}}H^{1.5}\ for\ H \le y \le h$
Velocity defect	$\frac{u_s - u}{u_s} = (1 - \frac{y}{H})^{1.5}\ for\ 0 \le y \le H$
Average velocity	$U = \frac{2}{3}\sqrt{\frac{\rho_m\, g\, \sin\theta}{\alpha}}H^{1.5}(1 - \frac{2}{5}\frac{H}{h}) = (1 - \frac{2}{5}\frac{H}{h})\, u_s$
Momentum correction factor	$\beta = (1 - \frac{11}{20}\frac{H}{h})(1 - \frac{2}{5}\frac{H}{h})^{-2}$

Table II Velocities for 2-D uniform debris flow in friction- collisional regime

2.4 Debris Flow in Macro-Viscous Regime

The existence of dilute suspended solid particles in a flow of granular-fluid mixture is believed to modify the flow of the fluid portion of the mixture thereby creating increased viscous dissipation. The relation of stress and strain rate is similar to that of Newtonian fluid until the volume sediment concentration exceed about 9%. Therefore, in this regime the granular-fluid mixture can be treated as a generalized Newtonian fluid with effective viscosity μ_m depending not only on the fluid property and temperature but also on the sediment concentration.

$$\tau = \mu_m \frac{du}{dy} \tag{9}$$

By calculating the increased viscous dissipation of dilute suspended spheres in a flow of a Newtonian fluid, Einstein (1956) theoretically obtained that $\mu_m = \mu_f(1 + 2.5\,C)$, where μ_f is fluid viscosity. Adding the effect of bound water (that is a thin layer of water adhering to the particle surface) on the effective sediment volume concentration, Chu (1983) obtained an extension of the Einstein equation, $\mu_m = \mu_f(1+2.5\,K\,C)$, where K is the ratio of the volume of sediment particles plus bound water to that of sediment particles. K is slightly larger than unity. The Einstein's equation and its extension have been proven to be valid only for very dilute spherical suspensions with negligible particle interactions. The modification should be in order when one desires to applied it to flows with heavily concentrated suspensions in which particle interactions are obviously significant. Assuming that the Einstein's equation is valid for an infinitesimal increase in the viscosity of the mixture in response to an infinitesimal increase in the concentration of particles, Krone (1984) obtained that $\mu_m = \mu_f \exp(2.5\,C)$. Chu (1983) considered the change of pore space due to particle collisions and obtained that $\mu_m = \mu_f(1+\varepsilon\,K\,C)^{-2.5}$, where ε is a coefficient that indicates the increase of intergranular pores due to particle collisions. ε is always larger than unity and it equals $4.5/\pi(\approx 1.43)$ for uniform spheres. Krieger and Dougherty (1986) obtained that $\mu_m = \mu_f(1 - K\,C)^{-B/K}$ with B quiet close to 2.5. Bagnold (1954) obtained that $\mu_m = 2.25\,\lambda^{1.5}\,\mu_f$ for flow in the macro-viscous regime from his rotating-drum experiments with low concentration of neutrally buoyant spheres. If the apparent viscosity μ_m is constant, the velocities of a steady two-dimensional uniform debris flow can be easily obtained and summarized in Table III.

Velocity distribution	$u = \frac{\rho_m\, g\, h^2 \sin\theta}{\mu_m}[\frac{y}{h} - \frac{1}{2}(\frac{y}{h})^2]$
Surface velocity	$u_s = \frac{1}{2}\frac{\rho_m\, g\, h^2 \sin\theta}{\mu_m}$
Velocity defect	$\frac{u_s - u}{u_s} = \left(1 - \frac{y}{h}\right)^2$
Average velocity	$U = \frac{1}{3}\frac{\rho_m\, g\, h^2 \sin\theta}{\mu_m} = \frac{2}{3}u_s$
Momentum correction factor	$\beta = 1.2$

Table III Velocities for 2-D uniform debris flow in macro-viscous regime

2.5 Debris Flow in Visco-Plastic Regime

The viscoplastic conceptualization of debris flows is founded largely on the idea that high concentration of sediment in flows increases viscosity and contributes shear strength to the flow. Shear strength in flows is provided by the fine-grained matrix which produces cohesion, and by the coarser particles which provide internal friction. Viscoplastic materials have a finite yield strength and flow as Newtonian fluid if the yield strength is exceeded.

$$\tau = \tau_y + \mu_b \frac{du}{dy} \tag{10}$$

Eq. 10 is called Bingham fluid model in which τ_y is Bingham yield stress (or shear strength) and μ_b is Bingham viscosity. The yield stress must be exceeded before flow occurs. Based on laboratory experiments and observations of natural debris flows, Johnson (1965) divided the shear strength into cohesion and friction parts and found that the friction part of the yield strength of debris-flow material is proportional to the normal stress σ acting on planes of shearing, and then he developed a model, called Coulomb-viscous model, having the form,

$$\tau = \tau_c + \sigma \tan\phi + \mu_b \frac{du}{dy} \tag{11}$$

The first two terms on the right hand side of the above equation represent shear strength $\tau_\gamma = (\tau_c + \sigma \tan\phi)$ that must be exceeded before debris flow occurs. The Coulomb- viscous and Bingham models are generally known as viscoplastic models. However, the values of the rheological parameters (τ_y and μ_b) are not invariant constants but vary with widely depending on the properties of the granular-fluid mixture such as solid concentration, clay type and cement, particle shape and its size distribution, temperature and electro-chemical properties of the liquid component in the mixture. For example, a recent experimental analysis of fine-grained slurries (with high solid volume concentration ranging from 0.44 to 0.66) conducted by Major & Pierson (1992) shows that yield strength and Bingham viscosity exhibit order-of-magnitude variation when sediment concentration changes as little as 2% to 4%. A method is required to compute the rheological parameters from the knowledge of relevant data of the mixture. Viscoplastic models are easy to apply, and they can explain some key features of debris flows, such as a rigid plug of relatively undeformed material rides along in the channel center and larger particles floating in the debris-flow matrix. Note that the macro-viscous and the viscoplastic models only describe the laminar mudflows. As the mudflows in turbulent situation, the turbulent stress must be considered that is proportional to the square of the shear rate. Table IV indicates the velocity characteristics of a steady 2-D uniform debris flow when τ_y and μ_b are constant.

Velocity distribution	$u = \frac{\rho_m\, g\, H^2 \sin\theta}{\mu_b}[\frac{y}{H} - \frac{1}{2}(\frac{y}{H})^2]$ *for* $0 \leq y \leq H$
Surface velocity	$u_s = \frac{\rho_m\, g\, H^2 \sin\theta}{2\mu_b}$ *for* $H \leq y \leq h$
Velocity defect	$\frac{u_s - u}{u_s} = (1 - \frac{y}{H})^2$ *for* $0 \leq y \leq H$
Average velocity	$U = \frac{\rho_m\, g\, H^2 \sin\theta}{\mu_b}\left(\frac{1}{2} - \frac{1}{6}\frac{H}{h}\right) = \left(1 - \frac{1}{3}\frac{H}{h}\right) u_s$
Momentum correction factor	$\beta = \left(1 - \frac{7}{15}\frac{H}{h}\right)\left(1 - \frac{1}{3}\frac{H}{h}\right)^{-2}$

Table IV Velocities for 2-D uniform debris flow in viscoplastic regime

2.6 Debris Flow in Visco-Plastic-Collisional Regime

In contrast to mudflows, stony debris flows contain a significant amount of larger particles as well as fine one, wider particle size distributions, and higher sediment concentration than mud flows. The interactions of larger particles as well as fluid viscosity may play significant roles in momentum exchange in flows. O'Brien & Julien (1985) proposed a physically based quadratic model that includes yield, viscous, collision, and turbulent stress components. That

$$\tau = \tau_y + \mu_d \frac{du}{dy} + (\mu_c + \mu_t) \left(\frac{du}{dy}\right)^2 \tag{12}$$

where μ_d =dynamic viscosity, μ_c =dispersive parameter (= $a_1 \rho_s \lambda^2 d^2$ defined by Bagnold) and μ_t=turbulent parameter (=$\rho_m l_m^2$, where ρ_m and l_m are the density and the mixing length of the mixture). The turbulent parameter is generally much less than dispersive parameter (Julien & Lan, 1991). When the dimensionless number D_ν [=the ratio of dispersive to viscous stresses(= $\rho_s \lambda^2 d^2 \mu^- 1_d (du/dy)$)] is less than 30, the model reduces to the simple Bingham plastic model. On the other hand, if D_ν is larger than 400, it reduces to the collision model. Considering the viscous effect may not coexist with the turbulent and granular dispersive effect in the same time, Chen (1988) developed a generalized viscoplastic fluid (GVF) model as

$$\tau = \tau_c \cos\phi + p \sin\phi + \mu_1, \left(\frac{du}{dy}\right)^\eta \tag{13}$$

where p=the dynamic pressure, μ_1 =consistency index, and η=flow behavior index. Chen's model can cover the spectrum of Newtonian, viscoplastic, dilatant, and power-law models depending how the yield stress, the consistency index, and flow behavior index are chosen. At this point it needs extensive experimental data to predict τ_c, ϕ, p, μ_1 and η. As mentioned by Chen (1988) himself, his model can reduce into Bingham model and Bagnold model when $\eta = 1$ and $\eta = 2$, respectively. When these coefficients are constant, the velocity characteristics of a steady 2-D uniform debris flow are summarized in Table V.

Velocity distribution	$u = \frac{\eta}{\eta+1}\left(\frac{\rho_m\, g\, H^{\eta+1}\, \sin\theta}{\mu_1}\right)^{\frac{1}{\eta}}[1-(1-\frac{y}{H})^{\frac{\eta+1}{\eta}}]$ $for\ 0 \le y \le H$
Surface velocity	$u_s = \frac{\eta}{\eta+1}\left(\frac{\rho_m\, g\, H^{\eta+1}\, \sin\theta}{\mu_1}\right)^{\frac{1}{\eta}}$ $for\ H \le y \le h$
Velocity defect	$\frac{u_s - u}{u_s} = (1-\frac{y}{H})^{\frac{\eta+1}{\eta}}$ $for\ 0 \le y \le H$
Average velocity	$U = \frac{\eta}{\eta+1}\left(\frac{\rho_m\, g\, H^{\eta+1}\, \sin\theta}{\mu_1}\right)^{\frac{1}{\eta}}(1-\frac{\eta}{2\eta+1}\frac{H}{h})$ $= \left(1-\frac{\eta}{2\eta+1}\frac{H}{h}\right)u_s$
Momentum correction factor	$\beta = \left(1-\frac{4\eta^2+3\eta}{6\eta^2+7\eta+2}\frac{H}{h}\right)\left(1-\frac{\eta}{2\eta+1}\frac{H}{h}\right)^{-2}$

Table V Velocities for 2-D uniform debris flow in visco-plastic-collisional regime, based on GVF model

3. Mixed-Layer Models

By regarding debris flow as a uniform mixture, various expressions of debris-flow velocity have been developed by adopting different models and summarized in Table I to Table V. The profiles of velocity distribution so obtained appear in the shape of a parabola. However, many experimental evidences (Takahashi, 1981; Savage, 1979; Yang and Wang, 1990; Shen, 1989/91) have indicated that the profiles of velocity distribution present a shape of an inverted "S" curve, rather than a simple parabolic curve. This inconsistency mainly results from the assumption of uniform debris-flow mixture. However, the distributions of sediment concentration are generally nonuniform due to the action of the gravity. The coefficients of the above mentioned models are strongly depend on the concentration and its distribution (Chen, 1988). Therefore, as the concentration varies with flow depth y, the coefficients of models are not constant but varies with y also. Except the above mentioned models, Su *et al.*(1993) proposed a mixed-layer model to simulate the velocity profile of debris flow under steady and uniform flow conditions. The mixed-layer model is deduced from an idea that one of the above mentioned model can not well model the debris-flow behavior throughout the entire depth. Therefore a combination of two or three of the models is proposed. Su *et al.* (1993) divided the debris flow into two layers. One is the macro-viscous layer which is near the bottom, and above the macro-viscous layer is inertial layer where the grain dispersive stress and turbulent shear stress are all important. After the comparison with experiments, they found that the mixed-layer model was better than other single-layer model in the simulation of velocity distribution. In contrary to Su *et al.*'s model, Yang and Wang (1991) proposed another mixed-layer model (called laminated layer by themselves). According to their experimental observations, they divided debris flow from the bottom to the free surface into three layers: bottom layer where the addition shear stress (resulting from particle contact and fluid viscosity) is dominant in the momentum exchange; inertial layer where the grain dispersive stress is dominant; and viscoplastic layer which lies upon the inertial layer and can be modeled by Bingham model. They also found that the velocity profiles obtained by using their mixed-layer model agreed closely with their experimental results. Based on these two examples, one can see that there is much research work to be done in this area.

4. Hydraulics of Debris Flows

Debris flows have been modeled as a Bingham substance, as a viscoplastic fluid, or as a dilatant fluid, as discussed in previous section. Use of one of these models for calculating velocities of debris flows requires estimation of coefficients related to shear strength or flow behavior. These coefficients have large ranges and considerable error is involved in their estimation. For example, Ling and others (1990) have pointed out that the flow-behavior index η and the consistency index μ_1 in the Chen's GVF model strongly depended on the flow depth, flow velocity, and particle concentration from their experimental study. Back-calculated from the simulated debris flows in the laboratory environment, the flow behavior index η varied from 0.25 to 2.05, and the consistency index μ_1 varied from 9.9 to 4415, with variation in three orders of magnitude. Such variation indicates that calculation of velocities from post-event evidence using these methods would involve considerable potential errors (Webb et al., 1987). Therefore some local empirical velocity equations are needed in the projects of mitigation of debris-flow disasters. Even though the hydraulics of debris flows are unusual compared with streamflow because of the high sediment concentration and the interactions among solid particles, some engineers regard debris flows as a special flow to be described by hydraulic formula, such as Manning equation or Chezy's equation, with modified roughness coefficients and exponents of hydraulic depth and friction slope. Empirical hydraulic formula of debris flow has the form of

$$U = m\,h^a\,S^b \tag{14}$$

where U is the average velocity; h is the flow depth; and S is the friction slope. The coefficient m and exponents a and b are to be determined in each field site. Different sites may have different values of the coefficients and exponents. As mentioned by Webb *et al.* (1987), there are other two methods in the estimation of average flow velocity. One is based on the elevation of velocity head and another on the superelevation on bends. Evidence for the elevation of velocity head usually is found where an obstacle is oriented perpendicular to the flow direction. Flow impinging on vertical walls will leave runup evidence in sites. The average velocity U is then calculated by equating the kinetic energy of the flow to the potential energy of the runup by

$$U = \sqrt{2\,\alpha_e\,g\,\Delta h_r} \tag{15}$$

where α_e is the energy-correction coefficient; g is the gravitational acceleration and Δh_r is the difference between the runup and the unobstructed flow-surface elevation. Another method considers that the surface profile of debris flow drops and rises on the inside and outside of a bend, respectively, to form an elevation difference Δh_s is due to the action of centrifugal-acceleration forces. The average velocity around the bend is estimated by

$$U = \frac{\sqrt{g\,R_c\,\Delta h_s}}{k\,W} \tag{16}$$

where R_c is the centerline radius of curvature; k is the correction factor and W is the effective channel width.

5. Summary

The present state of knowledge regarding debris-flow rheological models was reviewed. The flow velocities of a steady two-dimensional debris flow were obtained by using various models when debris flow was regarded as a uniform mixture in which the relating coefficients or parameters were considered constant. Since a single-layer model can not well describe the flow behavior throughout the flow depth in some cases, the mixed-layer model which consists of two or more models have been used by some investigators in order to fit their experimental velocity profiles. The coefficients, flow behavior index, or parameters of debris-flow models are strongly dependent on many factors, such as sediment size, sediment concentration and its distribution as well as the flow situation (in laminar or turbulent flow). To find the dependence of the coefficients or parameters on the above mentioned factors is one of the main research tasks in the near future. As debris flow is treated as a nonuniform mixture, a diffusion equation of debris-flow concentration is needed, except the mass and momentum conservation equations in solving the debris-flow behavior. Unfortunately there is little information about the diffusion mechanism and the deposition (or erosion) process of the debris-flow, and much research work is to be done in this area. Even though more attention should be paid to make the debris-flow mechanism clear, the goal of clearly understanding the debris flow may not reach in the near future due to the complexity of itself. Therefore it is still meaningful and important to investigate local empirical equations of mean velocity that are useful in the mitigation projects of debris- flow disasters.

REFERENCES

Bagnold, R. A. (1954), "Experiments on a gravity-free dispersion of large solid spheres in a Newtonian fluid under shear." Proc. R. Soc. London, Ser. A, 225, pp. 49-63.

Chen, L. C. (1988), "Generalized viscoplastic modeling of debris flow." J. Hydraulic Eng., Vol. 114, pp. 237-258.

Chu, J.(1983), "Basic characteristics of sediment-water mixture with hyperconcentration." Proc. 2nd Int'l Symp. on River Sedimentation, Nanjing, China, pp.265-273 (in Chinese).

Drucker, D. C. and W. Prager (1952), "Soil mechanics and plastic analysis or limit design." Q. Applied Maths. Vol. 10, pp. 157- 165.

Einstein, A.(1956), "Investigation on the theory of Brownian Movement." Dover Publications, Inc., New York.

Haff, P. K. (1983), "Grain flow as a fluid-mechanical phenomenon." J. Fluid Mechanics., Vol 134, pp. 401-430.

Hanes, D. M. and D. Inman (1985), "Observations of rapidly flowing granular-fluid materials." J. Fluid Mechanics, Vol. 150, 357-380.

Jan, C. D. (1992), "Movements of a sphere moving over smooth and rough inclines." Ph.D. Dissertation, Dept. of Civil Eng., University of California at Berkeley, USA.

Jenkins, J. T. and S. B. Savage (1983), "A theory for rapid flow of identical, smooth, nearly elastic, spherical particles." J. Fluid Mechanics, Vol. 130, pp. 186-202.

Johnson, P. C. and R. Jackson (1987), "Frictional-Collisional constitutive relations for granular materials, with application to plane shearing." J. Fluid Mechanics, Vol. 176, pp. 67-93.

Johnson, P. C., P. Nott, and R. Jackson (1990), ""Frictional- Collisional equations of motion for particulate flows and their application to chutes." J. Fluid Mech., Vol. 210, 510-535.

Julien, P. Y. and Y. Lan (1991), "Rheology of hyperconcentrations." J. Hydraulic Eng., ASCE, Vol. 107, pp. 346- 353.

Krieger, I. M. and T. J. Dougherty (1959), "A mechanism for non- Newtonian flow in suspensions of rigid spheres." Trans. Society of Rheology, Vol. 3, pp. 137-152.

Krone, R. B. (1984), "The significance of aggregate properties to transport processes.", Lecture Note on Estuarine Cohesive Sediment Dynamics, ed. by A. J. Mehta pp. 66-84.

Ling, C. H., C.L., Chen, and C.D. Jan (1990), "Rheological properties of simulated debris flows in the laboratory environment." Proc. Int'l. Symposium on hydraulics/hydrology of arid lands, ASCE, San Diego, CA, July 30-August 2, pp. 218-224.

Major J. J. and T. C. Pierson (1992), "Debris flow rheology: experimental analysis of fine-grained slurries." Water Resources Research, Vol. 28, No. 3, pp. 841-857.

McTigue, D. F. (1982), "A nonlinear constitutive model for granular material." J. Applied Mech. Trans., ASME, Vol. 49(6), pp. 291-296.

O.'Brien, J. S. and P. Y. Julien (1985), "Physical properties and mechanics of hyper-concentrated sediment flows, Proc. Specialty Conference on Delineation of Landslides, Flash Flood & Debris flow Hazards in Utah, pp. 260-279.

O'Brien, J. S. and P. Y. Julien (1988), "Laboratory analysis of mudflow properties." J. Hydraulic Engineering, Vol. 114, pp. 877- 887.

Pasquarell, G. C. and others (1988), "Collisional stress in granular flows: Bagnold revisited." J. Engingeering Mechanics, Vol. 114, pp. 49-64.

Pierson, T. C. and J. E. Costa (1984), "A rheological classification of subaerial sediment-water flows." Abstracts with Programs, 97th Annual Meeting GSA, Vol. 16(6), P. 623.

Prakash, J. R. and K. K. Rao (1991), "Steady compressible flow of cohesionless granular materials through a wedge-shaped bunker." J. Fluid Mech., V. 225, pp. 21-80.

Webb, R. H., P. T. Pringle and G. R. Rink (1987), "Debris flows from tributaries of the Colorado River, Grand Canyon national Park, Arizona." U.S. Geological Survey, Open-File Report 87-118.

Savage, S. B. (1979), "Gravity flow of cohesionless granular materials in chutes and channels." J. Fluid Mech., Vol. 92, pp. 53-96.

Savage, S. B. and S. McKeown (1983), "Shear stresses developed during rapid shear of dense concentrations of large spherical particles between concentric cylinders." J. Fluid Mechanics Vol. 127, pp. 453-472.

Savage, S. B. and M. Sayed (1984), "Stresses developed by dry cohesionless granular materials sheared in an annular shear cell." J. Fluid Mechanics, Vol. 142, pp. 391-430.

Savage, S. B. (1984), "The mechanics of rapid granular flows." Advances in Applied Mechanics, Vol. 24, pp. 289-366.

Shen, H. H. (1982), Constitutive Relationships for fluid-Sollid Mixtures, Ph.D. Thesis, Clarkson College of Technology, Postdam, USA.

Shen, H. H. and N. L. Ackermann (1982), "Constitutive relationships for fluid-solid mixtures." J. Engineering Mechanics Divsion, ASCE, pp. 748-763.

Shen, S. (1991), "The review and prospect about study of debris flow velocity." Proc. of Int'l Symposium on Debris Flow and Flood Disaster Protection, Emeishan, Sichuan, China, October 14-18, pp. 122-127.

Shen, S. and others (1989), "The experiment of velocity and concentration distribution of grannules in water-debris flow." Proc. 4th Int'l Symposium on river sedimentation, Beiging, China, pp. 649-656.

Spencer, A. J. M. (1964), "A theory of the kinematics of ideal soils under plane strain conditions." J. Mech. Phys. Solids, Vol. 12, pp. 337-351.

Spencer, A. J. M. (1982), "Deformation of an ideal granular material, in Mechanics of Solids." In: Mechanics of Solids (ed. by H. G. Hopkins & M. J. Sewell), pp. 607-652.

Su, C. G., H.P. Lien and Y.C.Chiang (1993), "Study on the velocity distribution of debris flow." J. Chinese Soil and Water Conservation, Vol. 24 (1), pp. 75-82 (in Chinese).

Takahashi, T. (1981), "Debris flow." Ann. Rev. Fluid Mechanics, Vol. 13, pp.57-77.

Takahashi, T. (1980), "Debris flow on prismatic open channel." J. Hydraulic Div., ASCE, Vol. 106, pp. 381-396.

Yang, M. and L. Wang (1991), "An experimental investigation on the laminated load model of debris flow." Proc. Int'l Symposium. on Debris Flow and Flood Disaster Protection, Emeishan, Sichuan, China, October 14-18, pp. 56-61.

DISCUSSION

Michiue: You reviewed many papers on debris flow. As the results, can you suggest which formula or theory is the most suitable for understanding the mechanics of debris flow?

Jan: A debris flow is a form of rapid mass movement of a body of of granular solids, water and air, with flow properties varying with water and clay content, sediment size , sediment sorting as well as flow speed. Therefore, the momentum exchange in debris flow is very complex. Due to the existence of solid particles in debris flow, the most suitable rheological model for understanding the mechanics of debris flow should include components to describe: (1) Cohesion between particles; (2) viscous interaction between particles and their surrounding fluid; (3) particle interactions, such as collision and friction; and (4) turbulence. O'Brien & Julien (1985/88) proposed a physically based quadratic model that linearly includes yield, viscous, collision, and turbulent stress components. That is

$$\tau = \tau_y + \mu_d \frac{du}{dy} + (\mu_c + \mu_t) \left(\frac{du}{dy}\right)^2$$

where τ_y = the yield stress including cohesion and friction stress; μ_d =dynamic viscosity; μ_c =dispersive parameter and μ_t =turbulent parameter. This model may be the most suitable formula for understanding the mechanics of debris flow up to now, but it is not a convenient one in application.

References

O'Brien, J. S. and P. Y. Julien (1985), "Physical properties and mechanics of hyper-concentrated sediment flows, Proc. Specialty Conference on Delineation of Landslides, Flash Flood & Debris flow Hazards in Utah, pp. 260-279.

O'Brien, J. S. and P. Y. Julien (1988), "Laboratory analysis of mudflow properties." J. Hydraulic Engineering, Vol. 114, pp. 877-887.

Davies: Experiments show that the velocity gradient du/dy is greatest near the bed of a grain-in-fluid flow, with average shear rate less than 10 1/sec. This means that the flow at the bed is more likely to be inertial and that the upper layers are more likely to be macroviscous. This is the opposition of the mixed-layer models you describe. However, field deposit of debris flows show no evidence of different layers.

Jan: Not all experiments show that the velocity gradient is greatest near the bed of a grain-in-fluid flow (i.e., parabolic velocity distribution). The experiments conducted by some investigators, such as Takahashi (1981), Savage (1979), Yang & Wang (1990), Shen (1989), Su *et al.* (1993) among others, have indicated that the profiles of velocity distribution present a shape of an inverted "S" curve, rather than a simple parabolic curve. That is to say that the greatest velocity gradient does not occur near the bed but somehow between the bed and the free surface, according the experimental results obtained by the above mentioned investigators. This inconsistency mainly results from the assumption of uniform debris-flow mixture. The velocity profile for debris flow with uniform distribution of concentration is more likely to be parabolic. However, the distributions of sediment concentration may be nonuniform due to the action of the gravity. The lower layer may have higher concentration than the upper layer. Higher concentration always results larger cohesion, friction and viscosity interaction in debris flow. Therefore, the velocity profile for debris flow with nonuniform distribution of concentration is more likely to present a shape of an inverted "S" curve. This gives some investigators, such as Su *et al.* (1993) and Yang & Wang (1990), motivations to develop multi-layer models. In addition, field deposit of a debris flow may show if there is sorting phenomenon or not, but can not show the sediment concentration distribution of an active debris flow is uniform or nonuniform.

References

Savage, S. B. (1979), "Gravity flow of cohesionless granular materials in chutes and channels." J. Fluid Mech., Vol. 92, pp. 53-96.

Shen, S. and others (1989), "The experiment of velocity and concentration distribution of grannules in water-debris flow." Proc. 4th Int'l Symposium on river sedimentation, Beiging, China, pp. 649-656.

Su, C. G. and others (1993), "Study on the velocity distribution of debris flow." J. Chinese Soil and Water Conservation, Vol. 24 (1), pp. 75-82 (in Chinese).

Takahashi, T. (1981), "Debris flow." Ann. Rev. Fluid Mechanics, Vol. 13, pp.57-77.

Yang, M. and L. Wang (1991), "An experimental investigation on the laminated load model of debris flow." Proc. Int'l Symposium. on Debris Flow and Flood Disaster Protection, Emeishan, Sichuan, China, October 14-18, pp. 56-61.

Dynamics of the Inertial and Viscous Debris Flows

Tamotsu Takahashi

Disaster Prevention Research Institute,
Kyoto University
Gokasho, Uji, Kyoto 611, Japan

1. Introduction

Debris flow may be defined as a highly concentrated flow of the mixture of water and sediments. The characteristics of the flow, however, are various in a wide spectrum of behaviors from very slow laminar to fast highly turbulent flow and from muddy liquid to stony sluggish flow corresponding to the properties of sediment materials as well as the hydraulic conditions such as velocity and depth. This suggests that a particular dynamic mechanism may exist for each typical regime of the flow. Qualitative classifications of the flow in various points of view have been proposed, but that in the light of the dynamic mechanism and the existence criteria of each regime are the themes of further investigation.

The evident difference in behaviors of the debris flow from those of the plain water flow or the thinnly sediment loaded flow may be brought by the effects of frequent encounters of particles and/or diminution of void among particles. Therefore, the main fundamental concerns in the mechanics of the debris flow would be the constitutive relations which properly take the roles of the densely concentrated sediments into account. In this context, attention should be paid to that the dominating factors among many possible relevant ones which control the stresses might be different case by case depending on the particle properties, concentration and the properties of the interstitial fluid.

2. Particle Sustaining Mechanism in a Plane Shearing Flow

When a granular material comprised of uniform spheres is most densely packed, the volume concentration of the solids in it, c, is as much as 0.741 ($c = c_{*0}$). The linear concentration, λ, defined by Bagnold [1954] is the ratio of the particle diameter to the mean free distance between the particles and it is given by the following equation.

$$\lambda = \{(c_{*0}/c)^{1/3} - 1\}^{-1} \tag{1}$$

The λ at the maximum possible concentration is, therefore, infinitive. The most sparse cannon ball packing is possible at $c_* = 0.605(\lambda \simeq 14)$, and the solids concentration in the most sparse square packing is attained at $c_* = 0.523(\lambda \simeq 8)$. At a solids concentration larger than $\lambda_2(\simeq 17)$, particles can not dislocate each other. General shearing of the granular material becomes possible when $\lambda < 17$, but if $\lambda > \lambda_3(\simeq 14)$, particles are always in touch with each other and the applying shear must overcome the resisting stress due to the internal friction and the yield strength of the interstitial fluid if any. When $\lambda < \lambda_3$, in some arrangements of particles, and when $\lambda < 8$, in any arrangement of particles, any particle is, on the average, free from other particles. The resistance to shearing in such a substance devoiding any skeleton structure should only originate from the resistance in the interstitial fluid (the viscous stress and yield strength if any).

Under the action of gravity, however, the particles heavier than the surrounding sheared fluid (no strength exists in the fluid under shear) cannot maintain their neutral positions but settle down to the bed. This means to keep the heavy particles dispersed in the entire depth, some particle supporting forces which balance with the submerged weight should act.

Provided the interstitial fluid exerts no major effect on particles other than buoyancy, the dispersive force should be attributed to the repulsion due to frequent collision of particles. Because only high relative encountering velocity can guarantee the particle dispersion by the inelastic collision, this mechanism seems to appear in the high speed regime. The inelastic collision necessarily consumes energy, so that, the shear stress is produced.

If the viscosity and density of the interstitial fluid are large, even in the slow flow regime, enough magnitude of dispersive force to suspend particles might be produced by the squeezing flow expelled from the void between the approaching particles. This squeezing flow dissipates energy and the excess shear stress will be produced.

Bagnold [1966] measured the already mentioned critical concentrations using natural beach sand whose diameters were between 0.318 and 0.414mm, and obtained; $c_{*0} = 0.644(\lambda = \infty)$, at the stable state after shear $c_{*1} = 0.604(\lambda_1 = 48)$, at the critical state beyond which shearing is impossible $c_{*2} = 0.555(\lambda_2 = 19)$, and at the least concentration beyond which particles are always in touch $c_3 = 0.51(\lambda_3 = 12.4)$. Those critical concentrations for the well graded mixture are not clear yet, but fine particles can enter into the void among the coarse particles and incidentally those concentrations may shift to larger values. For example, the samples obtained at the Jiangjia Gully in China showed $c_{*0} = 0.61 \sim 0.73$, whereas the very fast debris flows passed through the channel whose slope was only about 3° with the concentrations $c = 0.61 \sim 0.72$ (Wu et al. 1990). Content of very fine materials (smaller than about 0.05mm) seems important to determine the mechanics of debris flow, because they are not only effective to increase the apparent density of the interstitial fluid, but also increase the viscosity of the fluid and thereby influence the resistance to flow as well as particle dispersion pressure. Lubricating effect might also exist on the collision of coarse particles.

3. Constitutive Relations in Various Flow Regimes

3.1 Quasi-Static Regime

When velocity is very small, the dispersive pressure caused by particle collisions and the other dynamic stresses may be negligibly small. In such a flow, particles should be in close contact and sliding on one another. For approximately uniform natural sand this regime may only be able to arise under the solids concentration $c = 0.56 \sim 0.51$.

The constitutive equations for such a plane flow of submerged noncohesive particles in an open channel would be as following:

$$p' = (\sigma - \rho) g \cos\theta \int_z^h c dz \tag{2}$$

$$\tau = p' \tan\phi \tag{3}$$

where p': inter-granular pressure at height z measured from the bottom, τ: shear stress at height z, $\tan\phi$: internal friction coefficient, h: depth of flow, θ: channel slope, σ: density of particle, ρ: density of fluid, g: acceleration due to gravity. Because there is no reason for nonuniform distribution of c, (2) can be written as

$$p' = (\sigma - \rho) g \cos\theta c(h - z) \tag{4}$$

and then, the shear applying at height z is

$$\tau = \{(\sigma - \rho)c + \rho\}(h - z) g \sin\theta \tag{5}$$

The granular material may begin to flow when the condition $\tau > p' \tan\phi$ is satisfied. This conditon is obtained from (4) and (5) as

$$\tan\theta > \frac{c(\sigma-\rho)}{c(\sigma-\rho)+\rho}\tan\phi \tag{6}$$

From (6) by substituting $\sigma = 2.65$ g cm^{-3}, $\rho = 1.0$ g cm^{-3}, $c = 0.51$ and $\tan\phi = 0.8$, one obtains $\theta > 20°$. This means the quasi-static regime can only arise on a very steep slope, presumably in very beginning of a debris flow.

3.2 Macro-Viscous Flow Regime

Consider a case the volume concentration of large particles c in a slurry is less than 0.51, which means the individual large particles are, on the average, not in contact with one another. Even if the stationary slurry has a certain strength to sustain the large particles, once it is sheared and under continuous deformation, the skeleton structure in the slurry is already disappeared and no matrix strength remains. The source of the dispersive force to be able to sustain particles in such a flow should be attributed to the dynamic structure of the flow.

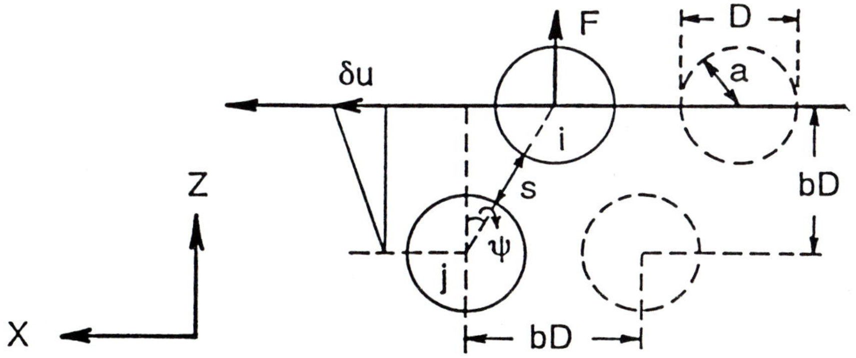

Fig. 1. Definition sketch of sheared granular material

Consider a particle i is moving with a relative velocity δu over a layer of particles as shown in Fig. 1, where the particles are immersed in the static viscous fluid. When the particle i approaches the particle j in the lower layer, fluid in between the two particles is expelled and a flow around the particle is generated. The access speed between the two particles in the direction of the line of connecting the centers is $\delta u \sin\psi$, and so the fluid dynamic force applying on the i particle opposite to the access direction is $6\pi\mu_f a^2 \delta u \sin\psi / s$, where μ_f is the fluid viscosity, a is the particle radius, ψ is the angle of the line of centers to the vertical direction at a certain time and s is the free distance between the particles [Davis et al. 1986]. The upward fluid dynamic force, F, acting on the i particle is then

$$F = 6\pi\mu_f a^2 \frac{b\sin\psi\cos^2\psi}{b-\cos\psi}\frac{du}{dz} \tag{7}$$

The i particle may exist at any position in the area of $4a^2b^2$, so that, the dispersive pressure applying to this area may be written as

$$p = \frac{3}{2}\pi\left[\frac{1}{b}\frac{\sin\psi\cos^2\psi}{b-\cos\psi}\right]_{av}\mu_f\frac{du}{dz} \tag{8}$$

where $[\,]_{av}$ means the mean value in the bracket in the process of approaching. As the particle i approaches to the j particle, the pressure increases, and at a certain ψ it becomes maximum and then decreases to zero at the top of the j particle, where $\psi = 0$. The mean value in the blacket during the total approaching process is larger, the smaller the b value, i.e., the larger the λ, the larger the dispersing pressure. This implies that to be capable to sustain a particle under a relative velocity, the distance between the particles should be smaller than a certain value, i.e., the concentration of particles should be larger than a certain value. As the i particle moves apart, a negative pressure may result by the backward flow to fill the gap between the particles. But this may not be as large as that in the approaching process due to weak return flow from larger area around the particle. The actual phenomena in the granular material should be more complicated, because the i particle is nothing but a randomly chosen one in the same layer it belongs to and the effects of assemblage of particles such as obstruction to the expelled flow due to narrow gap are neglected, when the i particle is going apart from the j particle, it is, at the same time, approaching the one standing in a line downstream of the j particle, and moreover, the particles in a layer do not form a straight line and the trace of the i particle approaching from lower position will form a curve around the j particle. Nevertheless, the dispersive pressure should be a function of solids concentration. Consequently, herein, the following formula is assumed.

$$p = g(\lambda)\mu_f \frac{du}{dz} = k\lambda^2 \mu_f \frac{du}{dz} \tag{9}$$

where k is a numerical constant.

The shearing stress would be the sum of the viscous stress in the interstitial fluid and that produced by the expelled flow.

$$\tau = \mu_T \frac{du}{dz} = f(\lambda)\mu_f \frac{du}{dz} + \Phi g(\lambda)\mu_f \frac{du}{dz} \tag{10}$$

where μ_T is the apparent viscosity of the flow and Φ is a numerical constant.

Because the particles cannot be deformed, the shear strain concentrates in the decreased space among particles, and this reduction of shearing space affects to increase the apparent viscosity of the interstitial fluid. If energy losses in the unit volume with and without particles are assumed equal, $f(\lambda)$ may be written as

$$f(\lambda) = 1 + \lambda \tag{11}$$

Then,

$$\frac{\tau}{p}(\equiv \Psi) = \frac{1 + \lambda + \Phi k\lambda^2}{k\lambda^2} \tag{12}$$

Bagnold [1954] gave the following formulae according to his own theory and experiments.

$$\mu_T = (1 + \lambda)(1 + 0.5\lambda)\mu_f \tag{13}$$

$$\Psi = 0.77 \tag{14}$$

The above mentioned discussions focused on the case where individual particles in the flow are not in touch with one another, and therefore, no direct stress transmission among the particles exists. When the solids concentration is large (for uniform particle case; $c > 0.51$), the effect of the particle contact becomes important. Even in such a case, if one watches a particular particle's motion, it would move in contact with other particles for a while and become free in another time. In this process, the total pressure p may be the sum of that directly transmitted between the particles, p', and the dispersive pressure produced by the expelled flow among particles. Consequently, it may be possible to flow on a flatter slope on which the quasi-static regime flow cannot exist.

3.3 Grain-Inertia Flow Regime

When the flow velocity is large and the interstitial fluid does not affect much, the constitutive relation should be mainly determined by the momentum transfer between colliding particles. Dimensional consideration leads to the following relationships:

$$p \,\text{or}\, \tau \propto \sigma D^2 \left(\frac{du}{dz}\right)^2 \tag{15}$$

where D is the diameter of particle. The coefficients of proportion may depend on the solids concentration, coefficient of restitution, and so on. Quite a few investigations have been done on such equations especially for the gas-solids phase granullar flow by analogy of the kinetic theory, and some theoretical and experimental investigations have treated the liquid-solids phase case.

Some results of physical model experiments as well as that obtained by a numerical experiment are shown in Fig. 2.

The ordinate f is the coefficient when τ is described as

$$\tau = f \sigma D^2 \left(\frac{du}{dz}\right)^2 \tag{16}$$

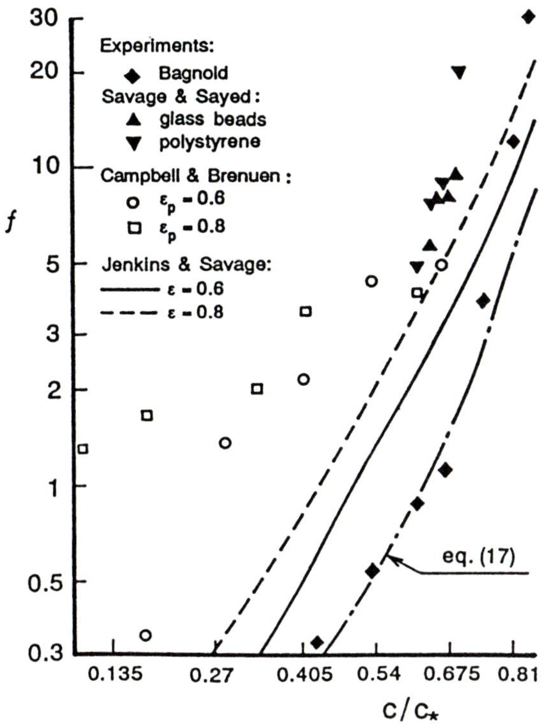

Fig. 2. The shear-stress function f versus solids concentration

Simultaneously shown in the same figure are the theoretical curve obtained by Jenkins and Savage [1983] and Takahashi's empirical relation [Takahashi 1991]

$$f = a_i \sin \alpha_i \lambda^2 \,, \ \tan \alpha_i = (1 + \lambda^{-1}) \tan \phi \tag{17}$$

where $a_i = 0.042$ and $\tan \phi = 0.65$ are substituted as representative values for spherical grains as those used in Bagnold's experiment.

The difference between the Bagnold's results and the other results may be caused, as Campbell and Brennen [1985] reasoned, by the difference in the experimental conditions; namely, Bagnold experimented the liquid-solids mixture flow in which wax spheres having small coefficient of restitution were used, whereas, the other experiments were for the gas-solids flow using nearly elastic grains. The other liquid-solids phase experiments such as Hanes and Inman's [1985] showed almost the same result to Bagnold's and that by Savage and McKeown [1983] showed larger f than Bagnold's. Daido et al. [1984] obtained one or two orders of magnitude less f than Bagnold's by using a similar apparatus to the latter. The reason for such large discrepancy was unknown, but in Daido's experiment there were large residual p and τ values under no deformation.

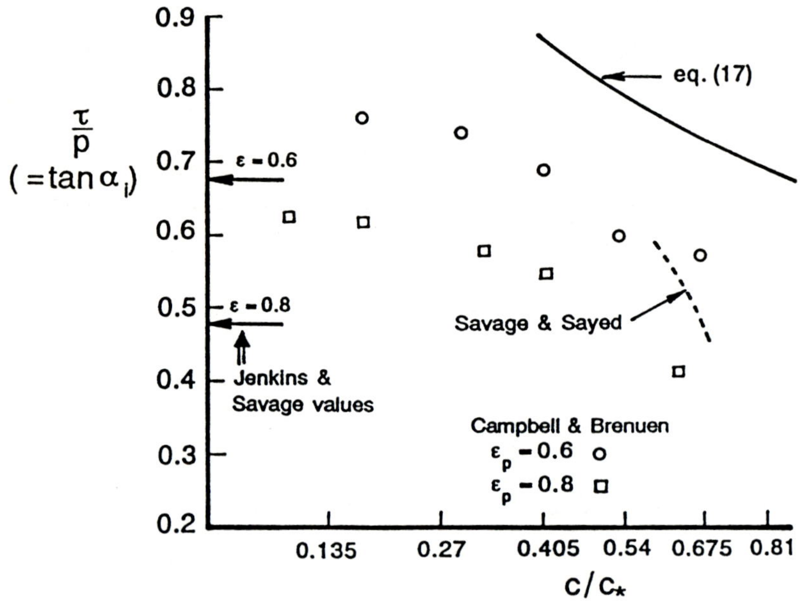

Fig. 3. The apparent friction coefficient as a function of solids concentration

Fig. 3 shows the change in $\tau/p(\simeq \tan\alpha_i)$ values versus solids concentration. The full line in the figure shows the second equation in (17). Because this line is drawn based on Bagnold's experiment, the absolute values are larger than those obtained by the gas-solids experiments. The tendency that $\tan\alpha_i$

becomes small with increase in solids concentration is, however, similar to the other cases. The cause of this tendency is explained by decrease in the angle of encounter between two particles with increase in solids concentration.

3.4 Turbulent Flow Regime

If the stresses due to collision of particles and turbulent mixing dominate among others, the shear stress may be written as

$$\tau = f\sigma D^2 \left(\frac{du}{dz}\right)^2 + \rho_a \ell^2 \left(\frac{du}{dz}\right)^2 \tag{18}$$

where ρ_a is the apparent fluid density and ℓ is the mixing length. In case the total applying shearing force is nearly balanced with the first term in the right hand side of (18) (grain-inertia regime), the mixing length would be at most the mean free distance among particles and ρ_a should be equal to the density of interstitial fluid. However, in case the solids concentration is low or particle diameter is small but the applying shearing force is large, the first term in the right hand side of (18) cannot be large enough to be able to balance with the total applying shearing stress and the remaining stress should be balanced with the second term. If the second term is far larger than the first term, particles in the flow are suspended by turbulence. In this case ℓ is larger than the mean free distance among particles and ρ_a is larger than that of interstitial fluid due to suspension of solids. This is the debris flow in the turbulent flow regime or the mud- flow type debris flow, and according to the open channel experiments using natural sand and water, this regime appears when $h/D > 20 \sim 30$ is satisfied [Arai and Takahashi, 1986]. In case the two terms in the right hand side of (18) have comparable magnitudes, a hybrid type flow may appear in which the grain-inertia regime arises in the lower part and the turbulent regime arises in the upper part of the flow. [Takahashi 1991]

3.5 Quadratic Constitutive Equations

It seems reasonable to describe the comprehensive flow regimes including the transition between the typical regimes by a pair of the quadratic constitutive equations. Various such propositions have been done under their own considerations. One typical expressions may be deduced from the above mentioned considerations:

$$\tau = \tau_y + \mu'_f \frac{du}{dz} + K\sigma D^2 \left(\frac{du}{dz}\right)^2 + \rho_a \ell^2 \left(\frac{du}{dz}\right)^2 \tag{19}$$

$$p = p' + \mu''_f \frac{du}{dz} + K'\sigma D^2 \left(\frac{du}{dz}\right)^2 \tag{20}$$

It would be clear already that all the terms in those equations cannot simultaneously be dominated in one particular flow regime or even in the transition. Therefore, otherwise one knows the characteristic variations of every terms in the wide spectrum of the flow regimes, these description cannot help a reliable prediction of the flow.

It may be noted here that some researchers claim two or more terms in the right hand sides of (19) and (20) are equally important in a particular flow regime. For example, if one adopts the constitutive equations of Daido et al. for grain-inertia regime, one finds that the sum of the third and fourth terms in the right hand side of (19) cannot balance with the applying shearing stress. This necessarily leads to the conclusion that the major part of the applying force is balanced by the first terms in the right hand side of (18) or (19); i.e., the yield stresses or the Coulomb type resistance [Egashira et al. 1989]. Attention should be paid, however, in such a concentration where the particles are on the average apart from one another, no strength inherent in a skeleton structure in the material can exist, and moreover, in the constitutive equation of Daido and others, the dynamic shearing stress and pressure due to collision of particles are less by more than one order of magnitude than those given by the experimental results of Bagnold and others and other theoretical constitutive equations. Their constitutive equation can explain their own experimental results by assuming the frequency of collision is described as, similar to Bagnold's consideration, $\delta u/(bD)$. This is equivalent to consider that collision between the two particular particles can occur only once when the upper one gets ahead the lower one. According to the experiments of Tsubaki et al. [1982], the angles and positions of encounter and dislodgeing on a particles are different. Although they reasoned that this occurred after a rubbing motion between the two particles, because those two particles are free from other particles and there is no reason to adhere to each other, it may be considered as well, some frequent collisions occur during an encounter of two particles due to vibration and rotation of not perfectly spherical particles. Energy loss by such a number of collisions should be larger than that arises by only one collision. This conjecture leads to a larger f than Daido's and may result in the stress balance without any Coulomb type resistance.

4. Characteristics of Flow in Macro-Viscous Regime

Consider a steady uniform flow in a rectangular channel, which is composed of highly viscous liquid and uniform particles and the solids concentration is less than 0.51. The pressure and the shearing stress balance equations are, respectively, from (9) and (10)

$$k\lambda^2\mu_f\frac{du}{dz} = (\sigma-\rho)g\cos\theta\int_z^h c dz \tag{21}$$

$$(1+\lambda+\Phi k\lambda^2)\mu_f\frac{du}{dz} = \frac{R}{h}\{(\sigma-\rho)g\sin\theta\int_z^h c dz + \rho g\sin\theta(h-z)\} \tag{22}$$

where R is the hydraulic radius and the shearing stress is assumed to distribute evenly on the bed as well as on the side walls.

Making some approximations and assuming $\Psi = const.$, the distribution of the solids concentration under the boundary condition; at $z = 0$, $c = c_b$, is obtained from (21) and (22) as

$$\left|\frac{c - c_\infty}{c}\right| = \left|\frac{c_b - c_\infty}{c_b}\right| \left(1 - \frac{z}{h}\right)^{3k\{(h/R)\Psi - \tan\theta\}} \tag{23}$$

where

$$c_\infty = \frac{\rho \tan\theta}{\{(h/R)\sigma - \rho\}(\Psi - \tan\theta)} \tag{24}$$

Equation (23) implies that if $c_b = c_\infty$, $c = c_\infty$ for the entire depth, if $c_b > c_\infty$, c decreases upward and at the surface $c = c_\infty$, and if $c_b < c_\infty$, the maximum concentration c_∞ appears at the surface and the concentration decreases downward. The experiments using a rigid bottom flume [Takahashi and Kobayashi 1993], however, show that flow without deposition can exists when the average particle concentration in the supplied flow, c_s, is smaller than or equal to c_∞ and the particles distribute in the entire flow depth only when c_s is between a certain critical concentration c_c and c_∞. When c_s is smaller than c_c particles can occupy only the lower part of the flow. In any case, particles distribute nearly uniformly throughout the particle mixture layer. In this context, Ψ in (24) is important and the experimental data suggest it is around 0.5. The value of c_c would be a function of $\mu_f, D, \sigma/\rho, du/dz$, etc. and in the experiments it was about 0.3.

When particles distribute uniformly througout the depth, the velocity distribution is obtained from (22) as following:

$$\frac{u}{u_*} = \frac{u_* h}{(1 + \lambda + \Phi k \lambda^2)\mu_f}\{(\sigma - \rho)c + \rho\}\left\{\frac{z}{h} - \frac{1}{2}\left(\frac{z}{h}\right)^2\right\} \tag{25}$$

where $u_* = \sqrt{gR\sin\theta}$. This is nothing but the parabolic velocity distribution for a laminar flow, in which the apparent viscosity μ_T is

$$\mu_T = (1 + \lambda + \Phi k \lambda^2)\mu_f \tag{26}$$

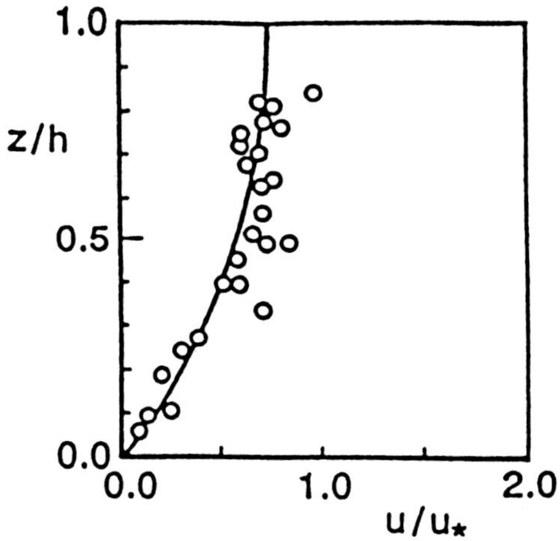

Fig. 4. Velocity distribution in the macro-viscous regime

Fig. 4 is an example of comparison of experimental velocity distribution with (25). The experiment was conducted in a rigid bed flume of 10cm in width and 16° in slope by introducing the mixture of sand ($D = 2 \sim 4$mm, $D_{50} = 3.25$mm, $\sigma = 2.65$g cm^{-3}) and the slurry made of kaolin powder and water ($\rho = 1.386$g cm^{-3}, $\mu_f = 1.5$ poise) whose sand particle concentration c_s was 0.287 (unit volume weight $\gamma = 1.74$g cm^{-3}). The depth of flow in this case was $h = 2.8$cm. To determine μ_T in (25), Φk should be known beforehand. The average value of it was obtained by experiments under various combinations of μ_f, c, θ and h. Although there are some problems to be further disclosed especially for very viscous slurry case, Φk seems to be nearly constant as long as μ_f is within a same order of magnitude. But if μ_f changes by an order or more, it seems drastically change.

Fig. 5 shows the specific viscosities versus particle concentration obtained by the experiments. This figure suggests that, in determining the apparent viscosity of the debris flow, when μ_f is comparably small, the effect of expelling flow working to disperse particles is more important than the effect of reduction of shearing space due to occupation by particles. When μ_f is very

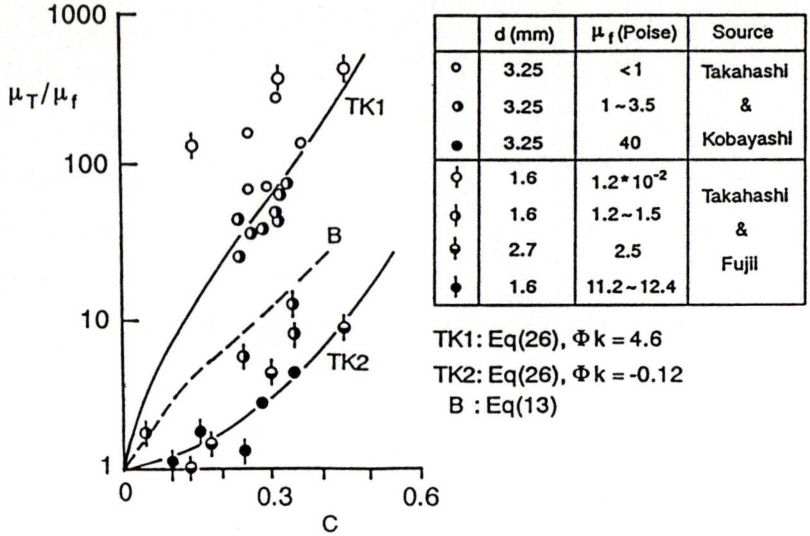

Fig. 5. Specific viscosities of the macro-viscous debris flows versus coarse particle concentration

large, the latter effect seems to become more important. The curve "B" in Fig. 5 is Bagnold's equation (13), and in this equation, the ratio of the two effects is always constant.

The magnitude of Ψ obtained in the experiments (around 0.5 in the author's experiments and 0.77 in Bagnold's experiments) is worth noting. Substitution of those values in (24) gives c_∞, which is the maximum possible particle concentration in the uniform equilibrium flow and if the channel slope is flat it becomes only a small value. Nevertheless, at the Jiangjia gully in China the debris flows are observed freighting particles by concentrations far larger than thus calculated equilibrium values. What mechanism is then working for such a high fluidity in the Jiangjia gully case? The key factor might be that the material is a well graded mixture. Provided particles in a certain grade of diameters are suspended in the expelling flow which is generated by shearing, the weight of those particles is transmitted to the fluid phase and it results in an increment of the aparent density of fluid. This works to increase buoyancy acting to larger particles and diminishes the necessary intensity of expelling flow to suspend the larger grade particles. Such a hierarchic particle supporting idea, although the fundamental mechanism considered is completely different, was first suggested by Rodine and Johnson [1976]. Therefore, the high fluidity may be the unique characteristics of the poorly sorted macro-viscous debris flow and to prove this further experimental investigations are needed.

5. Characteristics of Flow in Grain-Inertia Regime

The pressure and the shearing stress balance equations for a uniform flow are, respectively, from (15) and (17)

$$p = a_i \cos\alpha_i \lambda^2 \sigma D^2 \left(\frac{du}{dz}\right)^2 = (\sigma - \rho) g \cos\theta \int_z^h c\,dz \tag{27}$$

$$\tau = a_i \sin\alpha_i \lambda^2 \sigma D^2 \left(\frac{du}{dz}\right)^2 = g \sin\theta \int_z^h \{(\sigma - \rho)c + \rho\} dz \tag{28}$$

From those two equations and (17) with some approximations, the concentration distribution equation under the boundary condition; at $z = 0$, $c = c_*$, is obtained as following:

$$\frac{c}{c_*} = \frac{c_\infty / c_*}{(1 - c_\infty/c_*)(z/h) + c_\infty/c_*} \tag{29}$$

where

$$c_\infty = \frac{\rho \tan\theta}{(\sigma - \rho)(\tan\phi - \tan\theta)} \tag{30}$$

and this determines the maximum transportable concentration.

The distribution defined by (29) shows the maximum concentration is c_* at the bottom and it approaches c_∞ at the surface. Objection for using (27) and (28) even for the range of large concentration near the bottom may arise, because in such a large concentration, some parts of stresses may be transmitted as the static components and it seems reasonable to add the appropriate terms describing these components, respectively, to the left hand sides of (27) and (28). Fig. 2, however, suggests even in very large concentration, the effect of continuous particle contact is not large in comparison to the dynamic components. Therefore, herein, notwithstanding it contains evident contradiction in the neighborhood of zero velocity region, (27) and (28) are assumed to be satisfied in the entire flow region.

By substituting the concentration distribution (29) into the right hand side of (27), one obtains the following velocity distribution:

$$\frac{u}{\sqrt{gh}} = \frac{2A}{15}\left\{3\left(1-\frac{z}{h}\right)^{5/2} - 5\left(1-\frac{z}{h}\right)^{3/2} + 2\right\} \tag{31a}$$

$$A = \frac{1}{3}\left(\frac{\cos\theta}{a_i\cos\phi}\sigma - \rho\sigma\right)^{1/2} \frac{c_* - c_\infty}{c_\infty^{1/2}}\frac{h}{D} \tag{31b}$$

When $c = c_\infty$ at $z = 0$ is satisfied, the concentration is uniform throughout the depth, and the velocity distribution is given by

$$\frac{u}{\sqrt{gh}} = \frac{2h}{3D}\left[\frac{\sin\theta}{0.02}\left\{c_\infty + (1-c_\infty)\frac{\rho}{\sigma}\right\}\right]^{1/2}\left\{\left(\frac{c_*}{c_\infty}\right)^{1/3} - 1\right\}\left\{1 - \left(1-\frac{z}{h}\right)^{3/2}\right\} \tag{32}$$

Applicability of (32) for an inertia flow on a rigid bed have been proved by many experiments.

Examples of experimental velocity distributions on the movable bed are compared with (31) in Fig. 6 [Takahashi 1980]. To fit (31) in those examples, it was necessary to use larger a_i values than Bagnold's 0.042. This fact was used by some researchers to dispute the validity of the constitutive relations (16) and (17). The main reason for larger a_i values is, however, not inappropriateness of the constitutive equations but the effect of infiltration into the unsaturated bed. This was confirmed by the experiments having various degree of saturation in the movable bed and on the rigid non- pervious bed.

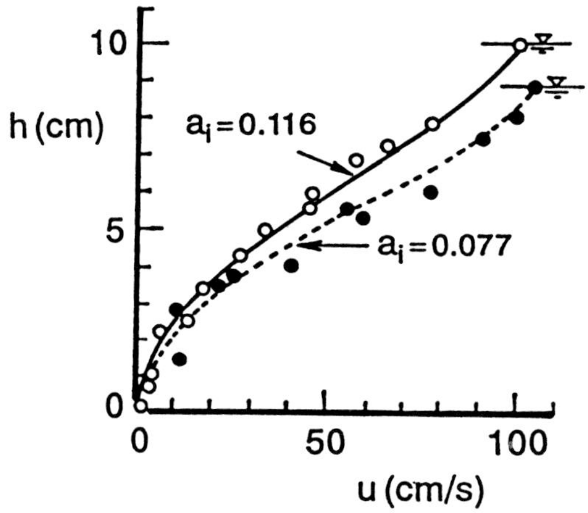

Fig. 6. Velocity distributions of the inertia-floe regime on a movable bed D=5.05mm, σ=2.65 g cm^{-3}, ρ=1.0 g cm^{-3}, $\phi = 36°$, $c_* = 0.65$. $\theta = 18°$

In the latter case, $a_i \sin \alpha_i = 0.02$ as used in (32) is a good approximation not only to predict the velocity distribution but also the magnitude of the velocity for an inertia flow of mixture of sand and water.

In a well graded mixture flow in inertia regime, if very fine fraction that suspends in the interstitial fluid is negligible, all the particles are sustained by repulsion. Due to imbalance between upward and downward repulsive forces and the effect of dynamic sieving, larger particles move upwards and transported downstream faster than the lower smaller particles. A routing procedure for the inertia debris flow taking such particle segregation process into account was recently given [Takahashi et al. 1992]. It must be noted that in the grain-inertia flow regime no buoyancy increase mechanism exists, so that, if big boulders are transported near the surface of the flow, they should be supported by frequent collison of smaller particles underneath.

6. Conclusion

The constitutive equations for respective flow regimes in the wide spectrum of debris flows were given considering the role of solids concentration, fluid viscosity, velocity, etc. based on the author's experiments and some viable theories and experiments checked by the critical reviews. Those constitutive equations were applied, respectively, to the macro-viscous regime flow and the grain-inertia regime flow, and the characteristics of those flows such as the equilibrium solids concentrations, velocities and their distributions are deduced. In the macro-viscous flow, the expelling force arising from approach of a particle to others plays the important roles to sustain particles and to determine the apparent viscosity of the flow. The particle concentration in an equilibrium macro-viscous flow composed of nearly uniform materials, in which the particles are transported without deposition, is not much different from that of the inertial flow on the same bed slope. This fact brings a kind of paradox that if the material is the well graded mixture, as observed in the actual viscous debris flow, the flow can transport much more dense solids concentration. A hierarchic buoyancy increment effect was suggested for a possibe cause of high competence to transport particles. Detailed investigation on the mechanics of macro-viscous flow of well graded mixture would be the key to disclose the whole aspects of the debris flows.

REFERENCES

Arai, M. and Takahashi, T. (1986) The mechanics of mud flow, Proc. JSCE, No.375, pp.69-77 (in Japanese).

Bagnold, R. A. (1954) Experiments on a gravity-free dispersion of large solid spheres in a Newtonian fluid under shear, Proc. Roy. Soc. London, A, 225, pp.49-63.

Bagnold, R. A. (1966) The shearing and dilatation of dry sand and the 'singing' mechanism, Proc. Roy. Soc. London, A, 295, pp.219-232.

Campbell, C. S. and Brennen, C. E. (1985) Computer simulation of granular shear flows, J. Fluid Mech., 151, pp.167-188.

Daido, A., Miyamoto, K., Miwa, H. and Nishimoto, N. (1984) A consideration on the constitutive equation of a granular flow containing Newtonian fluid in the void in view of energy loss mechanism, Proc. 39th Annual Meeting, JSCE, Part II, pp.367-368 (in Japanese).

Davis, R. H., Serayssol, J. M. and Hinch, E. J. (1986) The elastohydrodynamic collision of two spheres, J. Fluid Mech., 163, pp.479-497.

Egashira, S., Ashida, K., Yajima, H. and Takahama, J. (1989) Constitutive equations of debris flow, Annuals, D.P.R.I., Kyoto Univ., 32B-2, pp.487-501 (in Japanese).

Hanes, D. M. and Inman, D. L. (1985) Observations of rapidly flowing granular-fluid materials, J. Fluid Mech., 150, pp.357-380.

Jenkins, J. T. and Savage, S. B. (1983) A theory for the rapid flow of identical, smooth, nearly elastic particles, J. Fluid Mech., 130, pp.187-202.

Rodine, J. D. and Johnson, A. M. (1976) The ability of debris, heavily freighted with coarse clastic materials, to flow on gentle slopes, Sedimentology, 23, pp.213-234.

Savage, S. B. and McKeown, S. (1983) Shear stress developed during rapid shear of concentrated suspensions of larger spherical particles between concentric cylinders, J. Fluid Mech., 127, pp.453-472.

Takahashi, T. (1980) Debris flow on prismatic open channel, J. Hydraulics Div., ASCE, 106, No.3, pp.381-396.

Takahashi, T. (1991) Debris flow, Monograph Series of IAHR, Balkema, pp.1-165.

Takahashi, T. and Kobayashi, K. (1993) Mechanics of the viscous type debris flow, Annuals, D.P.R.I., Kyoto Univ., 36B-2, (in printing in Japanese).

Takahashi, T., Nakagawa, H., Harada, T. and Yamashiki, Y. (1992) Routing debris flows with particle segregation, J. Hydraulic Eng., ASCE, 118, No.11, pp.1490-1507.

Tsubaki, T., Hashimoto, H. and Suetsugu, T. (1982) Grain stresses and flow properties of debris flow, Proc. JSCE, 317, pp.70-91 (in Japanese)

Wu, J., Kang, Z. (1990) Observation research on debris flow in Jiangjia ravin, Yunnan,Beijing Science Press, pp.1-251 (in Chinese).

DISCUSSION

Davies: Why is the fluid dispersive pressure produced by expelling flow not symmetrical upwards and downwards? If the direction of motion of the grain i (Fig. 1) is reversed, Eq.(7) and Eq.(8) still apply but the sign of p is reversed. The streamline pattern is identical in both case, so the force on a body in the flow field is identical but of opposite sign.

Takahashi: Eqs.(7) and (8) show that F or p values have the same magnitude but having opposite signs provided ψ is the same when the i particle is approaching and going apart. This implies no net upward force applies to sustain particles in the shearing flow. However, Eq.(7) was deduced using Davis and others' analysis for approaching spheres and if the reverse flow intensity is not as strong as the expelling flow, Eq.(7) will not be held true in the going apart phase. This condition may be fulfilled if the expelling flow concentrates in a narrow space just adjacent to i particle and the reverse flow gathers from wider space. Difference of intensity between expelling and reverse flows may also produced if the magnitude of the acceleration in approaching phase is larger than in going apart phase. The trace of a particle in a layer may be meandering and the particle may repeat deceleration and acceleration.

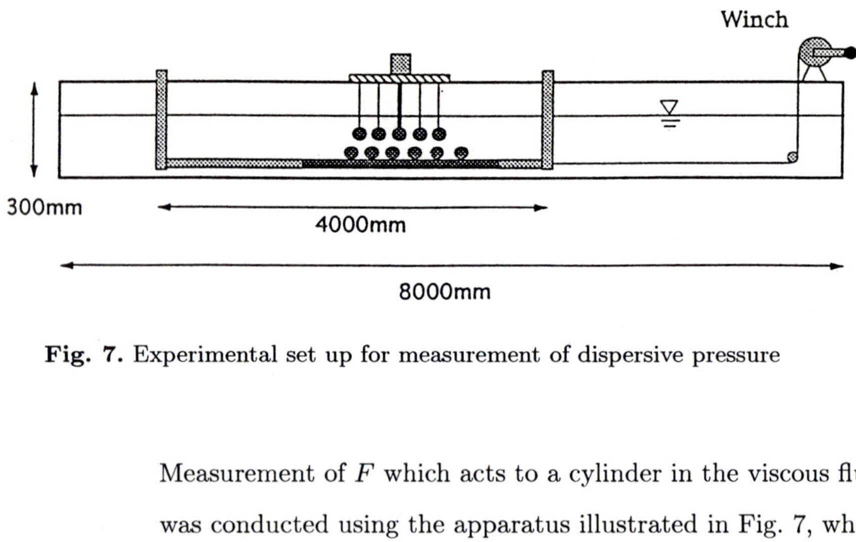

Fig. 7. Experimental set up for measurement of dispersive pressure

Measurement of F which acts to a cylinder in the viscous fluid was conducted using the apparatus illustrated in Fig. 7, where the lined up cylinders attached on the bottom were moved in a determined relative velocity to the cylinders in the upper layer. The vertical force applying to the cylinder in the center of the upper layer was measured. Fig. 8 shows the examples of net applying force in one cycle of rarative motion, where s is the gap between the upper and lower cylinders and μ is the viscosity of the solution of glycerol used in the experiment. Net upwards force was dedected in the experiments.

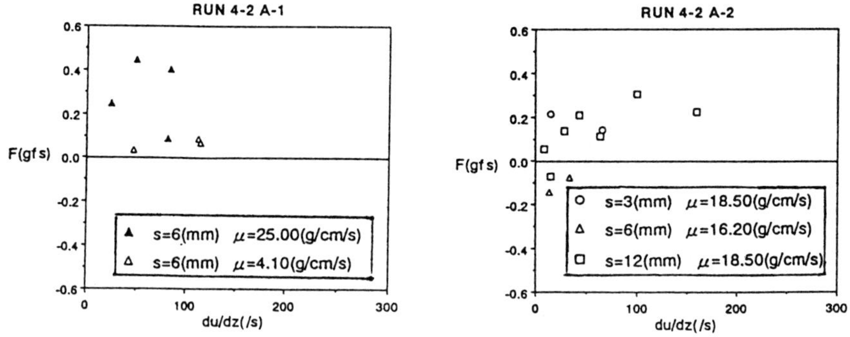

Fig. 8. The impulse applying the cylinder in one cycle of relative motion

Aguirre-Pe: Should the grain inertia regime be governed by the equations of grain inertia flow regime (Bagnold or Takahashi's equations) plus the turbulent shear stress?

Is it possible beforehand to predict if one is going to have stony debris flow?

Takahashi: The answer for the first question is "yes". As given in Eq.(18), shear stress in the grain inertia regime would be the sum of the dispersive stress due to grain collision and the shear stress in the interstitial fluid. But, in the grain inertia regime the mixing length ℓ is of the scale of mean free distance among the particles and ρ_a is equal to ρ. Comparison of the order of the two terms in Eq.(18) proves the second term is negligible.

To be the flow is macroviscous, the interstitial fluid should be highly viscous and therefore the sediment composition at the source area should contain with much clay and silt size fraction. Otherwise the flow becomes grain-inertia regime if it is possible to occur. Exact sediment size distribution pattern to divide the two regimes, however, is the problem to be disclosed.

Julien: 1. How do you calculate the dispersive pressure when $\sigma < \rho$ in Eq.(27)? 2. Is it correct to say that dispersive pressure reduces to 0 for neutrally buoyant particles ($\sigma = \rho$ in the right-hand side of Eq.(27))? 3. Can you elaborate on experimental evidence of the dispersive pressure in terms of experiments and references with experimental data?

Takahashi: 1. When $\sigma < \rho$, the particles tend to go upwards and by the existence of the ceiling (the free surface of the flow) they are restricted not to go beyond it. This means the inter-particle pressure increases upwards on the contrary to the case of heavier material. The absolute value of that pressure at height z from the bed may be given by

$$-(\sigma - \rho) g \cos\theta \int_0^z c dz$$

Therefore, Eq.(27) should be rewritten as

$$a_i \cos\alpha_i \lambda^2 D^2 \left(\frac{du}{dz}\right)^2 = -(\sigma - \rho) g \cos\theta \int_0^z c dz$$

It must be noted this equation applies only for the grain-inertia regime. The shear stress applying to the part near the free surface is small, so that in the actual flow the highly particle concentrated surface layer may flow like a plug with negligible velocity gradient. 2. Dispersive pressure should exist even in the case $\sigma = \rho$ as Bagnold confirmed by his experiment. In the case $\sigma > \rho$, the inter-particle pressure due to submerged weight of the particles may be balanced by the dispersive pressure as written in Eq.(27). But if $\sigma = \rho$, the dispersive stress by no means balance with the submerged weight of particles. Therefore, Eq.(27) does not make sense in this case. The dispersive pressure in the open channel flow with neutrally buoyant particles would play merely to increase the pressure in the flow. Actually, if the open channel flow with neutrally buoyant particles is experimented, the governing equation should be from Eq.(18)

$$\begin{aligned}(f\sigma D^2 + \rho_a \ell^2)\left(\frac{du}{dz}\right)^2 &= g\sin\theta \int_z^h \{(\sigma - \rho)c + \rho\}dz \\ &= \rho g \sin\theta (h - z)\end{aligned}$$

Because particles can be easily moved by the turbulence in the flow, ℓ can be much larger than D and consequently the velocity distribution becomes a log law type as usual turbulent flow of plain water. Fig. 9 shows an example of the experiment in which the polystyrene beads of $\sigma = 1.03\text{g/cm}^3$ and $D = 0.135\text{cm}$ is mixed to have $c = 0.38$ and supplied in a rectangular flume with $\theta = 6°$. The solid line on the figure is

$$\frac{u}{u_*} = 8.5 + 5.75 \log \left| \frac{z}{k_s} \right|$$

and in this case $k_s/D = 0.45$. The equivalent roughness seems to change with change in solids concentration as shown in Fig. 10. 3. Yes, I can. Many flume experiments confirm the equilibrium solids concentration in the inertial debris flow is often less than 0.5 (in volume). This means, on the average, particles are free from each other except the instant of collision, and because the turbulence in the interstitial fluid is not enough to suspend the coarse particle, those particles should be sustained by the effect of collision. This may be the proof of the existence and importance of the dispersive pressure. As for the viscous flow range, please examine the answer for the discussion by T. R. Davies.

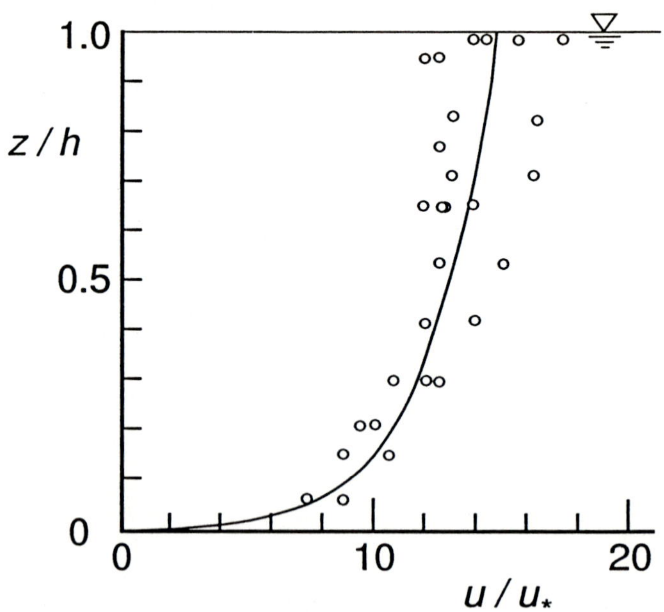

Fig. 9. Velocity distribution of the flow loading neutrally buoyant particles

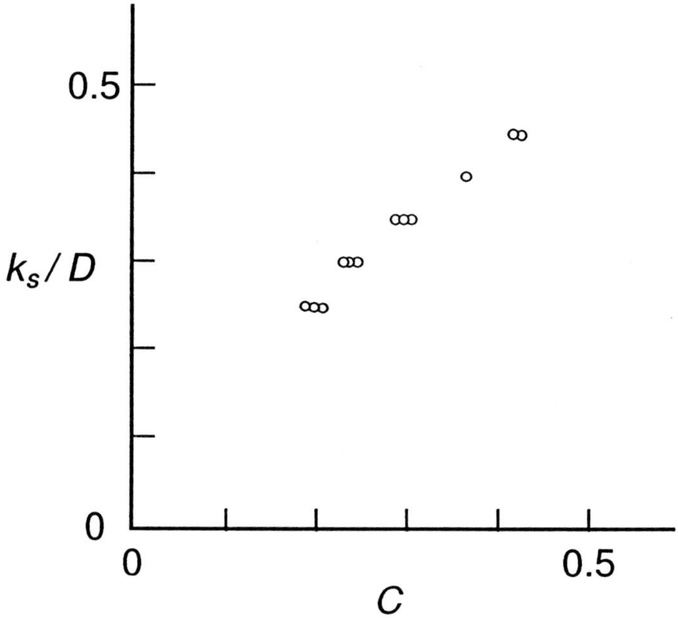

Fig. 10. Equivalent roughness versus solids concentration

Selected Notes on Debris Flow Dynamics

P.Y. Julien[1] and J.S. O'Brien[2]

[1] Engineering Research Center,
Colorado State University, Fort Collins,
CO 80523, USA

[2] Hydraulic Engineer, FLO Engineering, Inc.
P.O. Box 1659, Breckenridge,
CO 80424, USA.

Abstract

Heavily sediment-laden flows have been described and classified as hyperconcentrated sediment flows, including mud floods, mudflows, and debris flows. The authors prescribe definitions based on governing physical processes and limited concentrations of cohesive material. Viscous mudflows contain large concentrations of fine cohesive material. Rocky debris flows contain large concentrations of clastic material. Rheological analyses should recognize four types of shear stresses: 1) yield stress; 2) viscous stress; 3) turbulent stress; and 4) dispersive stress. These shear stresses combine into a quadratic rheological model. Dimensionless parameters from the ratio of shear stress terms identify the predominant physical process.

The two-dimensional model FLO-2D has been developed for the simulation of a wide range of hyperconcentrated sediment flows based on the quadratic rheological model. The simulation of the Pine Creek mudflow during the 1980 eruption of Mount St. Helens is presented as an example of our continuing progress in the physically-based analysis of natural disasters from heavily sediment-laden flows.

1. Introduction

The general classification of heavily sediment-laden flows describes various types of hyperconcentrated flows. Hyperconcentrated sediment flows ranging from water floods to debris flows are initiated with intense rainfall or snowmelt and may be triggered by hillslope and bank failures as well as landslides. Earthquakes and volcanic activities may also initiate the process of massive mobilization of liquefied soils in steep channels which may then deposit on alluvial fans. The flow properties and runout distances of these flow events are governed by the volume of the fluid matrix and the sediment properties.

Hyperconcentrations of non-cohesive particles with limited quantities of cohesive sediment display fluid characteristics at volumetric sediment concentrations $15\% < C_v < 40\%$ and are referred to as mud floods. Mud floods are turbulent and resistance to flow depends on boundary roughness. The sediment concentration tends to become fairly uniform throughout the flow depth because the increased fluid viscosity reduces the settling velocity of sediment particles. Woo et al. (1988) provided a detailed analysis of hyperconcentrations of sands.

In mudflows, the concentration of silts and clays is sufficiently high to bond the fluid matrix and to support clastic material. Mudflows behave as a singular fluid mass where boulders may be rafted along the surface. The fluid matrix has a relatively large concentration of sediments finer than 0.0625 mm and water. The volumetric sediment concentration of such fluid matrix roughly ranges from $45 - 55\%$ depending on the relative proportion of silts and clays. Mudflows exhibit high viscosity and high yield stress, can travel long distances on mild slopes at slow velocities and leave lobate deposits on alluvial fans. The flow is primarily laminar and local turbulence is quickly dampened. A detailed rheological analysis of mudflow properties has been presented by O'Brien and Julien (1988).

The analysis of debris flows stems largely from the contributions of Bagnold (1954) and Takahashi (1978). We suggest that debris flows represent a water-sediment mixture that contains significant quantities of boulders and debris where inter-particle impact is the dominant mechanism for energy dissipation. Debris-laden fronts may slow the progress of the flow or divert it in another direction. Particle interaction of sediment clasts can be a significant mechanism to transfer momentum to the flow boundary. Granular flows constitute a sub-class of debris flows in which the exchange of momentum between the flow core and the boundary occurs almost exclusively through particle collision. The water, which may be present in small quantities does not influence particle collision or lubricate the mass. Our understanding of sediment particle interaction in flowing water evolved from the study of O'Brien and Julien (1985). The definitions involving hyperconcentrated sediment flows should focus on the physical processes of the fluid motion which can be explored through the rheological study of sediment hyperconcentrations. Nomenclature has been formulated on the basis of what constitutes the fluid matrix (mixture of water and fine sediment particles) which govern the flow properties.

2. Rheology of Hyperconcentrated Sediment Flows

The non-Newtonian nature of hyperconcentrations results from several physical processes: The cohesion and bonding of fine sediment particles τ_c; the Mohr-Coulomb shear τ_{mc}, which is important when considering the static stability of steep slopes; the yield stress τ_y is defined as the sum of cohesive strength τ_c, plus the Mohr-Coulomb shear τ_{mc} and must be exceeded to initiate motion; the viscous shear stress τ_v which accounts for the increase in Newtonian viscosity; the turbulent shear stress τ_t which describes the turbulent nature of hyperconcentrated sediment flows of fine granular material; finally, the dispersive stress τ_d describes the effects of the collision of sediment clasts. Energy dissipation through turbulence, large eddies trailing major obstacles like trees and boulders, can be accounted for by considering τ_t.

The total shear stress τ in hyperconcentrated sediment flows includes contributions from each of these five shear stress components:

$$\tau = \tau_{mc} + \tau_c + \tau_v + \tau_t + \tau_d \tag{1}$$

When written in terms of shear rates, or velocity gradient $\left(\frac{du}{dy}\right)$, the following quadratic rheological model is obtained:

$$\tau = \tau_y + \eta \frac{du}{dy} + \zeta \left(\frac{du}{dy}\right)^2 \tag{2}$$

where

$$\tau_y = \tau_{mc} + \tau_c$$

$$\zeta = \rho_m l_m^2 + a_i \rho_s \lambda^2 d_s^2$$

In the above equations η is the dynamic viscosity of the mixture; τ_c is the cohesive yield strength; and τ_{mc} is the Mohr-Coulomb shear stress $\tau_{mc} = p_s \tan \phi$ depending on the intergranular pressure p_s and the angle of repose ϕ of the material; ζ is the inertial shear stress coefficient depending on the mass density of the mixture ρ_m, the Prandtl mixing length l_m, the sediment size d_s, the volumetric sediment concentration C_v, and ρ_s is the mass density of sediment. The mixing length l_m is usually given as a function of the distance from the boundary y and the von Karman constant κ . As a first approximation in depth-integrated flows, one can use the flow depth h, and a constant $\kappa = 0.4$ and the approximate mixing length is given by $l_m \cong 0.4h$. The coefficient a_i has been shown to vary widely and Takahashi proposed $a_i \cong 0.01$. Bagnold defined the linear sediment concentration λ as

$$\frac{1}{\lambda} = \left(\frac{C_m}{C_v}\right)^{1/3} - 1 \tag{3}$$

in which the maximum concentration of sediment particles $C_m \cong 0.615$. It is important to consider that the occurrence of debris flows as prescribed by a dispersive stress relationship alone requires that the following three conditions be simultaneously satisfied: 1) very large sediment concentrations, typically exceeding $C_v > 0.5$; 2) large velocity gradients typically exceeding $10s^-1$; and 3) very large grain sizes typically coarser than gravel in nature.

From equation 2, Julien and Lan (1991) proposed a dimensionless formulation of the quadratic rheological model in the form:

$$\tau^* = 1 + (1 + T_d^*)\, a_i D_v^* \tag{4}$$

in which the three dimensionless parameters $\tau*, D_v^*$ and T_d^* are defined as:

1. dimensionless excess shear stress τ^*

$$\tau^* = \frac{\tau - \tau_y}{\eta \frac{du}{dy}}$$

2. dimensionless dispersive-viscous ratio D_v^*

$$D_v^* = \frac{\rho_s \lambda^2 d_s^2}{\eta} \left(\frac{du}{dy} \right)$$

3. dimensionless turbulent-dispersive ratio T_d^*

$$T_d^* = \frac{\rho_m l_m^2}{a_i \rho_s \lambda^2 d_s^2}$$

It is suggested to relate the following parametric delineations to the classification of hyperconcentrations: 1) mudflows when yield and viscous stresses are dominant at $D_v^* < 30$; 2) debris flows or granular flows for which the dispersive stress is dominant at $D_v^* > 400$ and $T_d^* < 1$; and 3) mud floods when the turbulent shear stress is dominant at $D_v^* > 400$ and $T_d^* > 1$. A transition regime may be expected when $30 < D_v^* < 400$ for which all the terms of the quadratic equation are not negligible.

3. Two-Dimensional Simulation Model Flo-2d

Based on the quadratic rheological model, O'Brien et al. (1993) developed the two-dimensional flow routing model FLO-2D for the simulation of the continuum from water floods to mudflows. The momentum equation is solved after considering three components of the total friction slope S_f, namely: the yield slope Sy, the viscous slope S_v, and the turbulent-dispersive slope S_{td}. The total friction slope can therefore be rewritten as:

$$S_f = \frac{\tau_y}{\gamma_m h} + \frac{K\eta V}{8\gamma_m h^2} + \frac{n^2 V^2}{h^{4/3}} \tag{5}$$

in which γ_m is the specific weight of a mixture, h is the flow depth, V is the depth-averaged flow velocity, $K = 24$ for wide-rectangular channels but increases with roughness and irregular cross-section geometry, and n is Manning equivalent roughness coefficient for the turbulent-dispersive stress. The yield stress τ_y and the dynamic viscosity η increase with sediment concentration as defined by O'Brien and Julien (1988). The details pertaining to the model FLO-2D are available in O'Brien et al. (1993). Numerous mudflow hazard delineation projects have been completed using the FLO-2D model.

4. Mudflow Simulation of Pine Creek Using Flo-2D

In 1980, Mount St. Helens erupted, creating an explosive charge of gas, mud and water that cascaded as a pyroclastic surge down the cone of the volcano before collapsing into a high velocity mudflow or lahar down several drainages on the mountainside such as the Pine Creek channel sketched on Figure 1 (after Pierson, 1985). The data base of the Pine Creek mudflow was sufficiently complete to replicate the historic mudflow event. The mudflow traveled 22.5 km in 20 ± 3 minutes before entering Swift Reservoir where the mudflow volume and the peak discharge was estimated by response of a stage recorder at Swift Dam (Pierson, 1985).

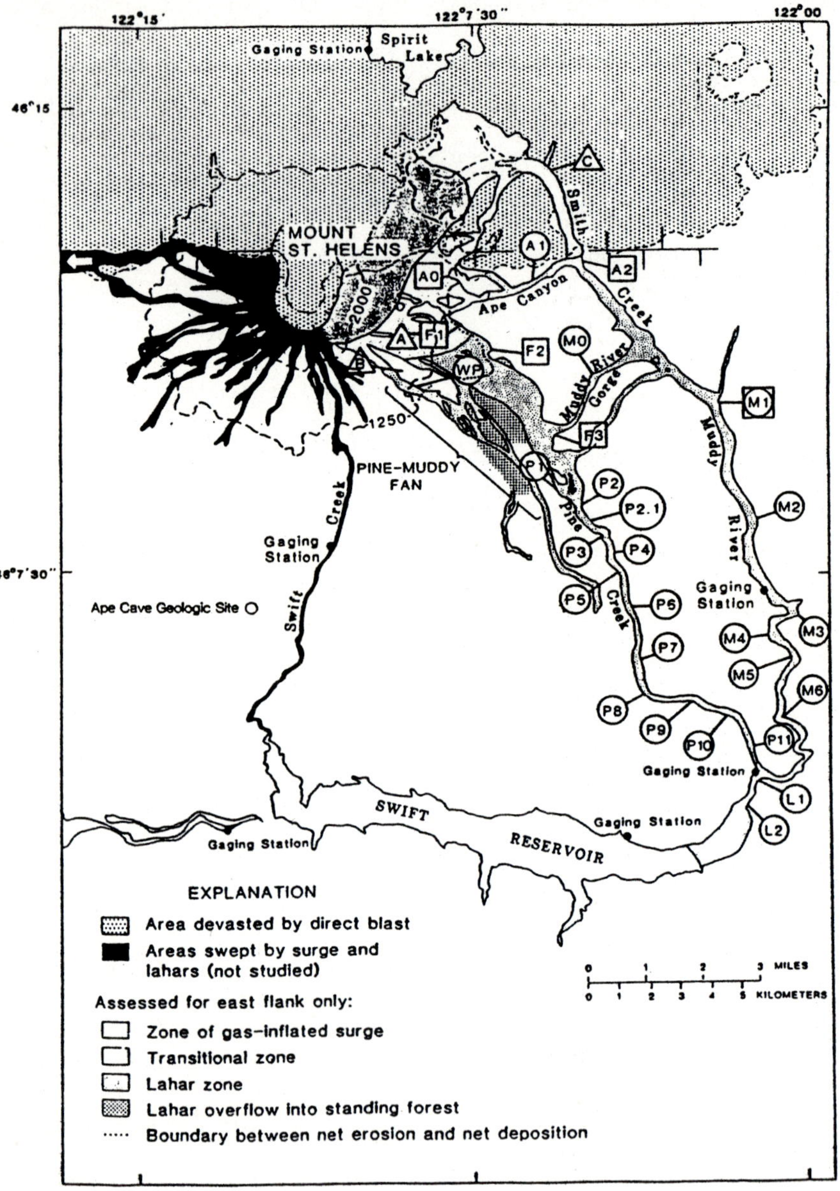

Fig. 1. Location of Pine Creek near Mount St. Helens (form Pierson, 1985; reproduced with permission of the Geol. Soc. of America)

It was necessary to estimate the initial flood hydrograph at the first cross-section to reproduce the volume of mudflow entering Swift Reservoir. The USGS provided cross-sections of Pine Creek surveyed after the event. It was reported that only minor amounts of overbank deposition and channel storage took place along Pine Creek.

The simulation of the Pine Creek mudflow was accomplished as follows:

1. a 15 minute topographic map was digitized and a uniform grid system of 500 ft square elements was established over the channel and potential flow areas;
2. a CAD program with a digital terrain model was used to export the grid element coordinates and elevations to a FLO-2D file;
3. rheological parameters for the mudflow were selected from Major and Pierson (1992) assuming a silt-clay to sand ratio of 1:1 for the Mount St. Helens mudflow. The viscosity and yield stress relationships as a function of volumetric sediment concentration were input parameters as a power regression;
4. the channel geometry data for 12 cross-sections were reduced and prepared in a data input file. Selected Manning n values ranged from 0.03 to 0.1. The distances between cross-sections for the FLO-2D simulation were approximated from the mapping provided by the USGS; and
5. the inflow hydrograph was estimated at the first cross-section to reproduce the estimated peak discharge at cross-section 2 and the inflow volume to the reservoir. The first grid element was located several thousand feet upstream of the first cross-section.

The computed mudflow viscosity and yield stress from Major and Pierson (1992) revealed that the equivalent sediment concentration ranged from 60 – 65%. FLO-2D was run several times to replicate the known flow conditions:

1. an estimated peak discharge at cross-section 2 equal to 28,600 $\frac{m^3}{s}$;
2. an estimated peak discharge at the reservoir of 7,500 $\frac{m^3}{s}$;
3. timing of the peak discharge arrival at the reservoir; and
4. estimated volumetric inflow to the reservoir.

When these conditions were satisfactorily met, the computed flow parameters were compared with those estimated by Pierson (1985) at 12 cross-sections. During the Pine Creek FLO-2D simulations, it was noted that an increase of 2% in sediment concentration would result in flow cessation on the falling limb of the hydrograph. This model response assisted in defining the limits in sediment concentration. There was still a question, however, whether the mudflow rheologic parameters used in the FLO-2D simulation would result in high velocity estimates for Pine Creek and would replicate the Swift Reservoir inflow hydrograph. The following FLO-2D results were obtained:

Hydrograph Timing - Arrival of the Peak discharge in Swift Reservoir

Pierson estimate: 20 ± 3 min. FLO-2D simulated: 20.4 min.

Volume - Total inflow volume into Swift Reservoir

Pierson estimate: 13,431,000 m^3 FLO-2D simulated: 13,490,000 m^3

Peak discharge - Peak discharge into Swift Reservoir

Pierson estimate: 7,500 m^3 FLO-2D simulated: 11,750 m^3

The hydrograph timing and volume from the FLO-2D simulations were the most accurate of the three flow conditions based on the response of the Swift Reservoir recording gage. The reservoir inflow peak discharge should be verified by reservoir routing and may be underestimated if the reservoir is shallow and floodwave attenuation was not accurately estimated in predicting the peak flow discharge.

Predicted peak flow velocity and depths correlated well with the estimated flow hydraulics from field data and the estimated values reported by Pierson (Table 1).

Table 1. Comparison of estimated and FLO-2D predicted flow hydraulics in Pine Creek

Xsection	Peak Disch. (m^3/s)		Max. Vel. (m/s)		Max. Flow Dep. (m)	
Grid	Pierson [1]	FLO-2D	Pierson	FLO-2D	Pierson	FLO-2D
P1-1166	17,100	28,300	23.5	21.1	9.8	20.6
P2-1043	28,600	27,200	17.7	20.9	15.2	19.2
P2.1-1009	25,900	27,000	20.8	21.8	12.6	14.0
P3-942	28,200	26,200	13.1	15.6	14.5	16.3
P4-915	21,700	25,000	12.4	19.4	14.9	18.2
P5-856	19,900	24,200	10.9	15.8	14.8	18.7
P6-672	21,000	21,700	14.2	14.0	13.9	20.2
P7-571	19,200	19,200	21.1	12.8	10.7	19.4
P8-432	16,600	18,000	15.3	12.7	9.4	13.0
P9-415	6,250	13,500	9.3	11.7	9.3	19.9
P10-372	8,930	12,500	11.0	9.6	9.0	14.6
P11-196	7,320	12,000	12.0	20.9	6.0	6.4

[1] Table 1 Average Flow Hydraulics from Pierson, GSA Bulletin, 1985, Vol. 96, p. 1064

The following considerations may explain some of the discrepancies. The flow depth was estimated from interpreted mudlines in the channel overbank areas and flow surging, cross-waves, variable cross-section geometry were assumed negligible in estimating the flow depth. It is likely therefore, that the flow depth was overestimated. The flow velocity was estimated by Pierson (1985) on the basis of the flow depth in the channel bends and a superelevation equation from which a peak discharge was computed. It follows that the velocity would be overestimated if the depth is overestimated. It should also be noted that the FLO-2D predicted maximum velocities and flow depths at a given cross-section do not necessarily occur at the same instant.

5. Conclusions

The rheology of hyperconcentrations is relatively complex, but the quadratic formulation appropriately describes the continuum of flow conditions ranging from mud floods to debris flows. The quadratic rheological model enables adequate two-dimensional computer simulations of yield, viscous, turbulent and dispersive stress in hyperconcentrated sediment mixtures.

FLO-2D is a flood routing model designed to simulate the continuum of water floods and mudflows in steep channels, over alluvial fans, and on urban floodplains. The Pine Creek mudflow triggered by the eruption of Mount St. Helens was properly simulated with the FLO-2D model. The relatively good correlation of the simulated results with estimated flow characteristics demonstrates the applicability of the model at volumetric sediment concentrations exceeding 60%. The analysis stresses the importance of appropriate values of rheologic parameters such as the dynamic viscosity and yield strength.

REFERENCES

Bagnold, R.A., 1954. Experiments on a Gravity-free Dispersion of Large Solid Spheres in a Newtonian Fluid under Shear, Proc. Royal Soc. of London, Ser. A, 225, 49-63.

Julien, P.Y. and Y.Q. Lan, 1991. On the Rheology of Hyperconcentrations, J. Hyd. Eng., ASCE, 117(3), 346-353.

Major, J.M. and T.C. Pierson, 1992. Debris Flow Rheology: Experimental Analysis of Fine-grained Slurries, Water Res. Res. 28(3), 841-857.

O'Brien, J.S., and P.Y. Julien, 1985. Physical Properties and Mechanics of Hyperconcentrated Sediment Flows, Proc. ASCE Hyd. Div. Spec. Conf on Delineation of Landslides, Flash Flood and Debris Flow Hazards, Logan Utah, June 1984, 260-279.

O'Brien, J.S., and P.Y. Julien, 1988. Laboratory Analysis of Mudflow Properties, J. Hyd. Eng., ASCE, 114(8), 877-887.

O'Brien, J.S., P.Y. Julien and W.T. Fullerton, 1993. Two-dimensional Water Flood and Mudflow Simulation, J. Hyd. Eng., ASCE, 119(2), 244-261.

Pierson, T.C., 1985. Initiation and Flow Behavior of the 1980 Pine Creek and Muddy River Lahars, Mount St. Helens, Washington, Geol. Soc. of America, Bull. V. 96, 1056-1069.

Takahashi, T., 1978. Mechanical Characteristics of Debris Flow, J. Hyd. Div., ASCE, 104, 1153-1169.

Woo, H.S., P.Y. Julien, and E.V. Richardson, 1988. Suspension of large concentrations of sands, J. Hyd. Eng., ASCE, 114(8), 888-898.

DISCUSSION

Armanini: 1. What is the role of roughness in debris flow? 2. What is the difference between debris flow, mudflow and mud flood?

Julien: 1. The question is interpreted to relate to channel boundary roughness, as opposed to surface roughness of individual grains and/or clasts. The influence of channel boundary roughness depends on the flow properties. By analogy with clear water flows, resistance to flow increases with boundary roughness in hydraulically rough turbulent flows but remains insignificant in either laminar flows or hydraulically smooth turbulent flows. Surface roughness should be dominant for turbulent mud floods. Resistance to flow in viscous mudflows depends primarily on fluid viscosity and surface roughness resistance should be small in mudflows. In more viscous flows, or in transition flows, some momentum may be transferred in bends and other flow direction changes such as flow around obstacles. Momentum flux of this nature may be attributed to roughness. 2. We attempted to quantify the relationship between these flow phenomena in previous writings (O'Brien and Julien, 1985). The nomenclature for debris flows is likely to remain muddled as long as the criteria for delineating debris flows, mudflows and mud floods are not quantitatively determined. We contend that better understanding can only be achieved through a rheological analysis of the fluid matrix. This is a complex problem in itself because fluid properties depend on sediment concentration and particle size distribution. The understanding gained from the quadratic rheological model in equation 2 of the paper is that the relative magnitude of the various shear stress components determines the flow type. The nomenclature should therefore depend on the relative magnitude of yield, viscous, turbulent and dispersive stresses. To specifically answer the question, debris flows characterize the motion of coarse granular material with particle impact generating dispersive stress without significant fluid shear stresses. In debris flows, particle impact is dominant everywhere, impact with boundary roughness elements is only part of the total resistance. Mudflows are very viscous to the extent that the entire flow is essentially laminar, resistance to flow depends on fluid properties as opposed to boundary roughness elements. Mud floods are turbulent and resistance to flow depends largely of surface roughness. In the case of mud floods, three types of roughness should be considered: 1) grain roughness over a plane surface; 2) roughness from large elements protruding into the flow such as bridge piers, and large obstructions including buildings and man-made structures; and 3) channel irregularities, sinuosity and changes in channel geometry. In summary, channel resistance should be dominant for mud floods, variable in debris flows and very small in mudflows.

Taniguchi: I find your paper interesting. You said that the volumetric concentration of a mudflow was between 45% and 55%. It think it is too high, and the speed of actual mudflows with such concentration is very slow from the results of my experiments. What do you think?

Julien: The classification in the 1984 paper only provides guidelines or approximate ranges of sediment concentrations expected for different types of hyperconcentrated flows from our field samples. With the work of O'Brien and Julien (1988), it became clearer that any classification based on sediment concentration alone is misleading. the concentration of fines (silts and clays) is most important to determine the yield stress and the viscosity of hyperconcentrations. For instance, it is possible to observe viscous laminar mudflows in laboratory flumes at volumetric sediment concentrations below 20%, but this requires the sediment to contain a larger proportion of clays than usually found in the field. Natural volumetric sediment concentration required for mudflows where the viscous stress largely overcomes turbulent and dispersive stresses usually corresponds to 45-55%. This question highlights the importance of carrying out rheological analyses based on samples representative of field conditions. It is relatively easy in the laboratory to repeat the rheological measurements under various sediment concentrations by controlling the amount of water mixed with the dried in-situ sample. It should be remembered, however, that even landslides, with volumetric concentrations exceeding 65%, can attain extreme velocities on steep slopes, provided that the content of fine sediment is very small.

Kitamura: Concerning the slope failure which you showed on the second slide: 1. Could you show the soil profile of the slope failure site? 2. What are the main factors to cause the slope failure?

Julien: 1. This particular field site showed shallow glacial soils on rough rock outcrops consisting of relatively friable sandstone. The material crumbles from the frequent freeze-thaw cycles during winter and early spring. 2. Slopes are inherently steep but stable when dry. Their stability depends primarily on the moisture content provided by rainstorms and snowmelt. The south-facing slope is subjected to rapid late-spring snowmelt that triggers slope failure. Field reconnaissance surveys should consider the amount of material readily available for transport in steep mountain gullies and available on watershed slopes. Instability indicators include tension cracks, steep loose material, bank caving, and poor vegetation.

Michiue: According to your explanation, in the quadratic terms of shear stress, the turbulent shear stress is dominant in the usual case of debris flow in comparison with the dispersive stress. But I think that the Prandtl mixing length will be influenced by the sediment concentration. It seems to be very difficult to distinguish these to components. What is your opinion on this point?

Julien: We refer to debris flows when the shear stress is dominated by particle impact, which besides dry avalanches and rock falls does not represent all types of hyperconcentrations. Mud floods on the other hand, exert shear stress primarily through turbulence. The question regarding the influence of sediment concentration on the mixing length is truly intriguing among academicians. Traditional understanding promoted by Vanoni, Ippen, and Chien among others showed that the mixing length $l_m = ky$ decreased with sediment concentration because k decreased to $k \cong 0.2$ at large concentrations from the clear water value $k = 0.4$. Coleman (IAHR, JHR, 19,2,1981) challenged this view by introducing a wake flow function while the von Karman constant remains at $k = 0.4$. Woo et al. (ASCE, JHE, 114,8,1988) combined viscous and turbulent stresses for detailed calculations of sediment concentration profiles. It remains unclear as to whether the mixing length varies or not with sediment concentration. In any event, a two-fold change in k from 0.4 to 0.2 is very small compared to the thousand-fold variability of yield and viscous stresses in hyperconcentrations. It can be assumed as a first approximation that $k = 0.4$ for all practical applications to hyperconcentrations and debris flow.

Aguirre Pe: The mixing length $l_m$in $\zeta = \rho_m l_m^2 + a_i \rho^2 d_s^2$ and in $T_d^* = \rho_m l_m^2 / (a_i \rho_s \lambda^2 d_s^2)$ is very important to classify the type of flow and therefore to use the appropriate friction slope in calculations of debris flow. Could you give us some insight about how to determine the mixing length?

Julien: This a very important practical question. As per the response to the previous question, the issue awaits theoretical developments. Until significant research suggests otherwise, there is little evidence that the von Karman varies by more than a factor of 2, which is small compared to the uncertainty in evaluating the viscosity and yield strength of debris flows. For this reason, simple calculations based on $k = 0.4$ are recommended for all calculations. This value is sufficiently accurate at low concentrations. At high concentrations, the other terms of the quadratic equation are usually larger than the turbulent stress.

Takahashi: 1. How did you divide the matrix and the coarse materials in the natural samples? 2. In your simulation of mudflow deposition, what was the condition to stop the flow? I suppose you needed the strength value for the whole materials including coarse materials. 3. Can you sustain particles in suspension with viscous stress only?

Julien: 1. The fluid matrix contains all particle sizes finer than 0.0625 mm. The concentration of silts and clays ($d_m < 0.0625$ mm) defines the sediment concentration of the fluid matrix which is used to determine the yield stress and the viscosity of the hyperconcentration. 2. The condition that stops the flow is determined by the yield stress including the Mohr-Coulomb strength. Only the fine fractions (fluid matrix) determine this strength for mudflows. In the case of debris flows without fines (silt and clay), flow stoppage would be determined by the Mohr-Coulomb criterion only. 3. In mudflows, yield strength alone is sufficient to maintain small particles in suspension without settling if the grain size is smaller than:

$$d_{sb} = \frac{3\pi}{2}\frac{\tau_y}{\lambda_s - \lambda_m}$$

where τ_y is the yield strength, λ_s is the specific weight of sediment and λ_m is the specific weight of the mixture. Very coarse clastic particles settle in mudflows at a velocity which is largely reduced because of the large viscosity of the water-sediment mixture. The pressure gradient at the wavefront also contributes to move large boulders in the downstream direction. In turbulent mud floods, sand particles can be maintained in suspension by turbulence alone. In summary, at low concentrations, turbulence sustains particles in suspension. As sediment concentration increases, the yield stress also contributes to maintain suspension.

Davies: 1. What are the design criteria for debris flow/mudflow/mud flood protection structures: a) How do you select the size of the structure? b) What procedures do you have to cope with the super-design event when it happens? 2. You showed a slide of a mudflow with a dense bouldery front, and stated that the coarse grains were not important to the phenomenon. This is clearly nonsense, because if you remove the boulders the phenomenon changes fundamentally.

Julien: 1. a) Debris flows generally involve large inertial impact forces with clastic material. It is usually advisable to build sabo dams with thick concrete-walled structures along with ways to drain the interstitial fluid. Without fluid, debris flows rapidly come to a halt. Mudflows are quite different in that the velocities are slow, depths are large and volumes are relatively limited. It is often rewarding to guide mudflows to predetermined storage areas where oozing mudflows can come to a halt. The deposit areas can thereafter be excavated by machinery and the storage capacity replenished after each event. Mud floods must be treated nearly like regular turbulent flows. Effective methods include reducing boundary roughness by channel straightening, obstruction removal and channel lining. Containment berms can be built on floodways to induce sediment storage on the floodplain. For detailed structural design, the flow depth, velocity and impact forces at specific locations can be calculated with the aid of models such as FLO-2D once the magnitude of the flow event and the properties of the fluid-sediment mixture have been determined. b) Public regulations, zoning and development avoidance on alluvial fans is often indicated. Public awareness is possible at all times. For instance, communities developing on alluvial fans can be informed anytime about preventive measures, insurance and community improvement plans. At the onset of large events, people should be informed, kept away or evacuated from potentially hazardous areas prone to possible structural failure. 2. It is clear that in mudflows with bouldery fronts, the boulders do not generate fluid motion, but it is rather because of the high viscosity of the fluid matrix that boulders can be carried downstream. If you remove the boulders, the fluid maintains its viscous properties and fluid motion is relatively unchanged. The statement was therefore correct in that the coarse grains do not significantly alter the fluid properties of viscosity and yield strength which control the flow condition. The effect of coarse grains on fluid flow is therefore not so important.

Chapter 3

Control Measures for Debris Flows

Introduction

Aronne Armanini

Debris flow dynamics is a relatively new scientific branch; however, the problem of defence from debris flows is quite old and for many centuries several types of protection structures have been developed in different countries.

However, such kind of structures have often been constructed on the base of the designer's personal experiences rather than on the base of real scientific assumptions. In many countries, specific institutions are devoted to the problem of preventing debris flow damages. In these countries, where the problem is of great importance especially for safety, real plans against debris flow occurrences are drawn up.

The structures against debris flow are often divided into two categories: *active works* and *passive works. Active* countermeasures are those devoted to reduce the debris feeding. These works are generally designed to increase the stability of debris deposits. They are based on water drainage systems able to reduce the water content in debris deposits, on forestation and on slope consolidation works, where possible. *Passive* structures, like check dams, training dikes, stilling basins and others, are usually inserted in the torrents, in order to control the flow of debris. Design criteria for such a kind of works are lacking, but very often also the working conditions are unknown. The most important parameters necessary to the correct design are those of kinematic type: debris flow velocity, maximum discharge or hydrograph and volume.

In order to estimate the project hydrograph many attempts to link the debris flow discharge to the liquid discharge have been presented in the literature [see Introduction to Chapter 1, by Shimokawa]. A very recent approach has been presented also in this book [Hirano]. Once the hydrograph and the maximum discharge have been estimated, it is necessary to evaluate the debris flow velocity and the flow cross-section. The knowledge of the velocity and the discharge of debris flow project event are necessary to design debris flow channels: these calculations are normally made under the assumption of local uniform flow conditions. For this purpose, a rheological approach to debris flow dynamics is necessary. Different theories, based on the characteristics of the materials involved, have been proposed. Among them, many are derived from Bagnold dispersive theory, or from similar theories for non-homogeneous fluids. This attempt was first successfully developed by Takahashi[1991], who reports also in the present book some new specifications of his approach. A good review of these theories is reported here by Deng Jan and Shen, while new approaches are presented by Hashimoto, who compares flows of dry sand and sand-water mixtures, and by Davies, who ascribes the debris flow characteristics to the channel geometry. Other formula are based on laboratory or field observations (see for example Julien and O'Brien for the role of cohesion).

Control works of debris flows are usually inserted along the channel for different purposes: to reduce the velocity of the flow, to reduce its erosion capacity, to reduce the discharge, to deviate the flow direction, etc.. The most frequent structure is the check dam. A great variety of forms and structures have been designed to control ordinary sediment transport and massive sediment transport in torrents [Armanini et alii, 1991]. A wide review of such works adopted by Japanese torrent control agencies is reported by Okubo et alii in this volume. Particular attention is paid to the slit dam and to the steel cell dam. This is a special open check dam developed in Japan in order to control debris flows.

Unlike the ordinary check dam, the check dam design to control debris flows is subject to a very large hydrodynamic pressure, so that very often these structures are very strong. The problem of debris flow dynamic impact is tackled by Armanini in his paper, both from a theoretical and from an experimental point of view: two kinds of dynamic impact have been observed depending on mixture characteristics of debris flow.

Finally, a very interesting problem of bottom erosion amplification during debris flows is presented by Jäggi and Pellandini: proper suggestions suitable to avoid check dam collapse due to erosion amplification in case of debris flow are offered in the paper.

REFERENCES

Armanini A., Dellagiacoma F. and Ferrari L. (1991): From the check dam to the development of functional check dams, Lecture Notes in Earth Sciences n. 37, Springer-Verlag Berlin.

Takahashi T. (1991): Debris flow, IAHR Monograph Series, A.A. Balkema, Rotterdam

Development of New Methods for Countermeasures against Debris Flows

Shun Okubo, Hiroshi Ikeya[1], Yoshiharu Ishikawa and Takashi Yamada[2]

[1] Sediment Control Department
Ministry of Construction, 2-1-3, Kasumigaseki,
Chiyoda-ku Tokyo, Japan

[2] Erosion Control Department
Public works Research Institute
Ministry of Construction, 1 Asahi
Tsukuba City, Ibaraki Prefecture, Japan

Abstract

The principal characteristics of the disasters due to debris flows are heavy casualties and the time taken for restoration after major sedimentation. Damage from debris flows can be prevented or minimized with "hardware" measures such as erosion control facilities. A variety of such facilities has been developed to meet differing requirements such as preventing flows or controlling factors like generation, movement and deposition, and many have proved successful in field exploration tests. In recent years steel has been increasingly used due to its workability and functionality. Two examples presently undergoing testing and improvement are: the permeable type, or grid dam, designed to boost the capturing volume of a dam by allowing small and medium freshets sediment discharge to pass in preparation for debacle; and Debris flow breaker screens, which accelerate deposition by separating water from debris flows.

With ordinary slit dams, the slit width is varied so as to catch boulders and smooth out ordinary levels of sediment discharge. To provide for the possibility of sudden major sediment discharge, for instance during the second half of a debacle, research is underway into sabo dams with gates at the slit which can be opened and closed to control sediment discharge. Field testing of new types of sabo dams is still in progress, and a number of unresolved issues remain with respect to planning, design, construction and maintenance.

1. Introduction

Of the variety of natural disasters seen in Japan debris flows account for a great many deaths. Prevention and minimization of damage from debris flows is thus an issue of not inconsiderable importance, and a large number of universities and national research institutes are currently engaged in research into the mechanisms behind generation, movement and deposition, as well as prediction and evacuation systems. The Ministry of Construction and prefectural governments are using the findings in development of strategies and countermeasures.

Debris flow countermeasures can be classified as "hardware" (for instance erosion control facilities) and "software" (warning and evacuation systems, land use regulations). Both types are combined together to produce effective countermeasures for preventing and reducing damage from debris flows.

Development of hardware, considered fundamental to debris flow strategies, is focused on aspects such as scale and type of flow and purpose of facility. Examples of facilities already built at field sites includes sabo dams, slit dams, debris flow breaker screens, large drainage conduit dams, debris-flow dispersing and depositing areas, settling basins, dispersing forest zones, training dikes and training channels.

Software countermeasures are required during construction of hardware, or even after completion if it is considered that the hardware does not provides enough protection.

This paper outlines the basic principles behind debris flow countermeasures, and describes some current issues and newly developed strategies.

2. About Debris Flows

Japan began to tackle the problem of debris flow after the 1966 disaster at Lake Saiko in Yamanashi Prefecture. The pace of experimental and theoretical research quickened, with more field observations and deposition surveys to clarify the mechanisms behind the phenomena of generation, movement and deposition. Lately the hazardous torrents prone to debris flows* has increased even in suburban districts due to the advancing urban sprawl and the accompanying concentration of population in cities and land development and utilization in urban areas. According to a survey by the Ministry of Construction's Sabo Department, 79,318 hazardous torrents prone to debris flows throughout the nation are currently under threat.

Damage from debris flows is often a result of not just the size of the particles and speed of flow but also their nature of occurring unexpectedly. The 1975 flow at Mt. Iwaki in Aomori Prefecture damaged 28 houses (93% of the 30 houses in the local village) and took the lives of 49 people (70% of the village population of 70). The 1966 Lake Saiko flow destroyed nearly all the houses in the village, forcing the inhabitants to relocate (see Fig. 1).

Name of disaster	Prefecture	Region	Date	Death toll (% of village)	Properties damaged (% of village)
Mt. Iwaki	Aomori	Kurasukezawa	Aug 75	70% 49	93% 28
Shodoshima	Kagawa	Tanijiri	Sept 76	26% 32	50% 22
Nagasaki	Nagasaki	Toppo-mizudaira	July 82	57% 33	85% 17
Kaito	Mie	Ohara	Aug 82	43% 9	100% 7
Ohtaki	Nagano	Takikoshi	Sep 94	38% 16	100% 19

Fig. 1. Examples of destruction caused by debris flows [1]

Hazard torrents prone to debris flows are common in the mountains of Japan, and heavy rain can indirectly cause serious damage by triggering a flow. One example is Shodoshima Island where, in July 1974, debris flows

* defined as the hazard torrents prone to debris flows with the potential to cause damage to at least five dwellings or a government or public office, school, hospital, railway station, hotel or power generation facility.

developed in 70% of the 18 rivers considered dangerous, threatening Uchinomicho in the Yasuda region. Another is the 1982 Nagasaki disaster, where flows were observed in 70% of the 141 torrents designated as hazardous from debris flows.

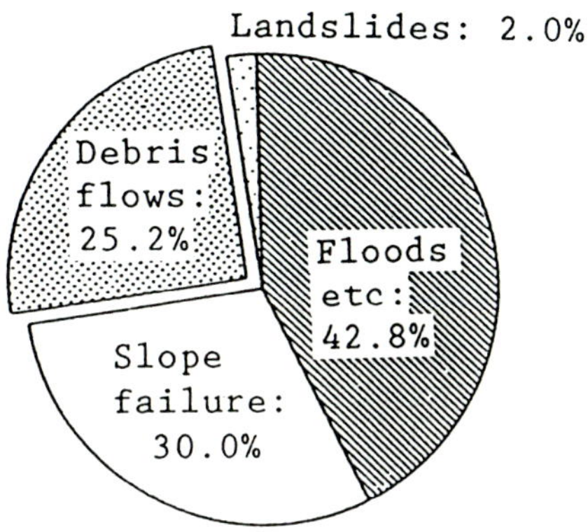

Debris flows:	1161
Landslides:	93
Slope failure:	1378
Floods etc:	1971
Total:	4603

Fig. 2. Statistics for dead and missing by phenomenon [2]

Figure 2 shows the numbers of dead and missing for each phenomenon during the period 1967-1981 [2]. At 60% sediment-related disasters account for more than half the total, and of this 25% (1161 dead or missing) is due to debris flows.

The 10-year average rate of disasters caused by debris flows between fiscal 1983 and 1992 was 839, and the number of dead and missing 14. The number of debris flows resulting in deaths for the same period was 4 with and average of 4 [3] people dead or missing in each.

3. Basic Principles Behind Countermeasures

Debris flow facilities can be classified into:

1. works for restricting the occurrence of debris flow
2. debris-flow capturing works
3. debris-flow direction controlling works
4. debris-flow training dykes
5. debris-flow dispersing forest zone
6. debris-flow depositing works.

Common practice is to us a combination of multiple units or different strategies, since single facilities are usually insufficient. The facility size and strategy used will vary according to local topographic conditions, feasibility of construction, cost, importance of the river basin and external forces. Fig. 3 shows a typical debris flow facility [4].

1. Works for restricting the occurrence of debris flow

Debris flows can be caused by fluidization of riverbed sediments or sediments from mountain slope collapse, or when a landslide dam formed by the latter type of sediments is destroyed by river water. Prevention strategies must take all three possibilities into consideration. Past experiments [5] indicate that small check dams, for instance, cannot suppress growth at the front of debris flows.

Preventing the movement of deposits on torrent beds is central to controlling the generation of debris flows in the first place. Sabo dams, check dams, consolidation, piling and hillside work are some of the ways to prevent such movement.

2. Debris-low capturing works

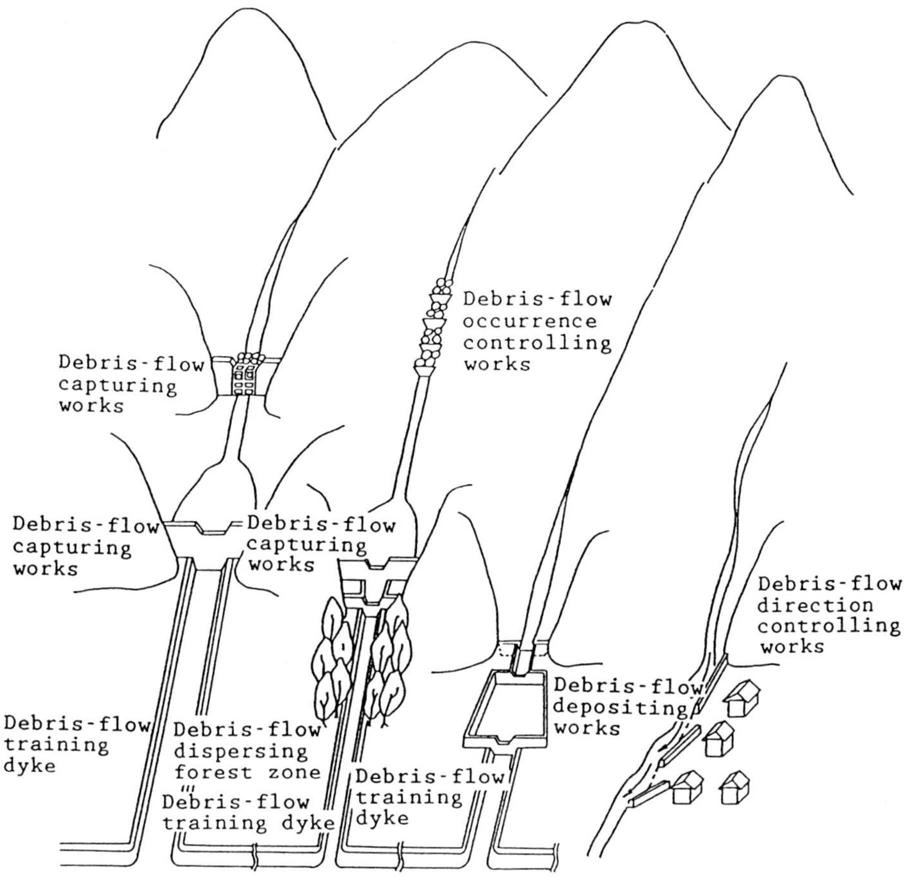

Fig. 3. A typical example of debris-flow countermeasure facilities [4]

A typical example would be sabo dams, both impermeable types such as ordinary sabo dams, and permeable types such as slit, debris flow breaker screen and large drainage conduit dams.

These can:

- capture debris flow so as to reduce the volume of sediment discharge
- lengthen the period of time from the occurrence of debris flow to its outflow onto alluvial fan

- prevent the movement of deposits on torrent bed (It is expected to be accomplished at the foundation section of permeable dams)
- capture the boulders and driftwood at the front of debris flow
- turn the debris flow into sediment flow
- reduce the peak discharge of debris flow

Permeable dams such as slit dams are in addition required to maintain a sedimentation capacity for the next debris flow by having no sedimentation from small and medium freshets, and the effect of a culvert boom to stop the driftwood which runs out.

3. Debris-flow direction controlling works

Debris flows can be guided to safe place via excavated training channels with cross-sections large enough to handle peak flow discharge levels. These channels are usually connected to one or more capturing dams or deposition facilities. Since debris flows tend to move straight forward the shape of alignment must be as straight as possible or, where curves are unavoidable, the radius of curve must be as large as possible. In order to prevent sediment accumulation from reducing the cross-sectional area of a river both the upstream movement of deposits from a confluence with the main river on from a point of change in gradient, and the need to avoid acute changes in the gradient of the longitudinal alighment, should be taken into account.

4. Debris-flow training dykes

The direction of a debris flow can be controlled by a training dike high enough to prevent overflow.

5. Debris-flow dispersing forest zone

Debris-flow dispersing forest zones, used to control the direction of movement as well as quantity of deposition by debris flows in alluvial fans, are commonly combined with work in 4) above. They can also be used as buffers between the descending flow and the object to be protected. Further research is need into the dynamics of impact forces in debris-flow dispersing forest zones.

6. Debris-flow depositing works

The energy of the flow is dissipated and deposition is encouraged using settling basins (for deposition in channels) or Debris-flow Dispersing and Depositing Areas (at outlet of gorge). Particularly at the foot of active volcanoes, where the volume of sediment discharge is such that deposition below is inevitable, Debris-flow Dispersing and Depositing Areas with enough space for deposition are required and sediment must be excavated systematically after deposition.

4. Examples of Countermeasures

This section briefly describes steel-made sabo structures and slit dams, both of which have become increasingly popular in recent years.

4.1 Steel-Made Sabo Structures

Steel-made sabo structures have excellent workability and cost performance in many respects including flexibility, permeation of sediments and water, short working periods, high quality, labor-saving, minimal transport facilities and associated costs and the ability to continue with construction during snow or cold periods. As a result a large number have been built throughout Japan, especially since the late 1960s.

Steel-made sabo structures are classified in terms of storage and permeability of sediment discharge in small and medium freshets as either permeable or impermeable. For planning reasons some of the latter type can pass water but not discharged sediments.

Of the impermeable dams, screen structures pass water but store sediments and are thus suitable where there is a plentiful supply of spring water or where constant water storage is undesirable, while the flexibility and water permeability of frame structures suit them towards volcanic regions and areas prone to landslides. The latter are also suitable for use in emergency disaster facilities because of their high workability.

Permeable dams were developed in order to boost sediment storage capacity during flooding by causing small and medium freshet sediment discharge to descend below the dam or hastening the cessation of flow and deposition of sediments in debris flow by separeting them from the flow.

4.1.1 Steel Slit Dams and Grid Type Steel-Made Sabo Structures. Both steel slit dams and grid type steel-made sabo structures consist either partly or wholly of permeable structures to boost capturing volume in preparation for debacles by allowing small and medium freshet discharge to pass. Debris flows stored and the energy dissipated (or water and sediments separated), thus arresting the flow. These dams can be used to store and regulate sediment discharge in rivers where the volume of sediment discharge in normal freshets is low and degradation below the dam is pronounced. Further they can serve as booms for driftwood.

Steel slit dams consist of A-type steel-pipe frames anchored in concrete at intervals of 2 - 4 metres. The A-frame is reinforced by filling the steel pipes with concrete. A recent development has been a B-type solid frame which maintains stability even when the angle of collision deviates from the direction of the center of stream.

Grid type steel-made sabo structures have a solid grid of steel pipes, a multilayer, multispan rigid frame structure with tightly welded contact points and a concrete foundation (see Photo 1). This technique can be used in building major dams.

Photo 1 Grid Type Steel-made sabo structures at Otanazawa on tributary of Komu River, branch of Fuji river

4.1.2 Debris Flow Breaker Screens. These sabo structures slow the movement of a debris flow, encouraging deposition and eventually bringing it to a stop using duckboard-style steel members forming a screen shape on the riverbed, and separating water from debris-flow while the flow is descending.

The impacts of boulders collisions in rapid flows exert tremendous force on ordinary gravity sabo dams, slit dams and erosion control facilities designed to arrest flow. Debris flow breaker screens should be capable of arresting a debris flow by dissipating its energy without being directly subjected to such forces (see Photo 2).

Debris flow breaker screens consist of a screen section (superstructure), supporting structure (sub-structure) and sidewall section to regulate flow movement, as illustrated in Fig. 4.

Photo 2 Debris flow captured by debris flow breaker screen at Kamikamihorisawa, Mt. Yakedake

Debris flow breaker screens were used on the Furano River, Mt. Tokachi following successful hydraulic model test by Watanabe [6] et all and follow-up

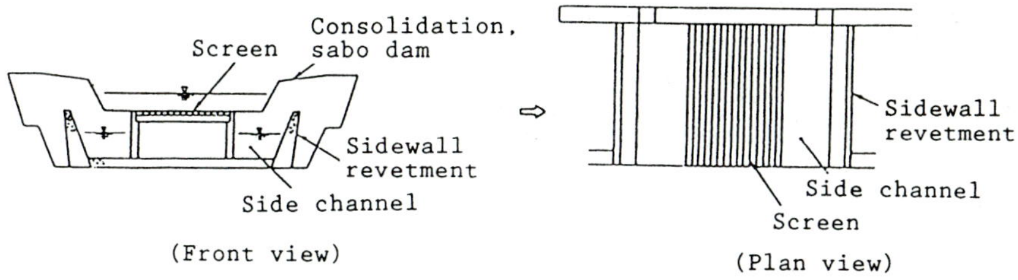

Fig. 4. Example of debris flow breaker screens

field tests at Iwadoi (Mt. Fuji), Kamikamihorisawa (Mt. Yakedake) and the Nojiri River.

Past surveys and tests of performance of debris flow breaker screens revealed that:

- spacing between screens is required to prevent blinding in a normal freshet for muddy debris flows (such as those in active volcanic areas) containing highly concentrated fine particles (such as volcanic ash). At Sakurajima Island and Mt. Tokachi, for example, the width used for horizontal screens was 40cm, or 90% of the size of the riverbed deposit particles.
- Deposition of sediments is accelerated if the following flow is prevented from spreading laterally. Further, deposited sediments will be pushed downriver if the lateral flow of a following flow is intercepted by sidewalls. At Sakurajima this has been avoided by providing a cross-section for the descent of the past maximum discharge without a sidewall on the right bank, thereby enabling the flow to fall freely from the screens. At Mt. Tokachi the structure is designed to guide the following flow through a side channel via a double section for the notch of the regular dam.

At Mt. Yakedake screens were installed on sidewalls as well, significantly accelerating the process of deposition and completely arresting the front of deposition (as can clearly be seen from VTR analysis and deposit surveys). It was thus established that the combination of side and bottom screens achieves faster deposition than bottom screens alone.

The cross-sectional form of screen members was also studied in an attempt to find ways of reducing screen blinding and removing deposited sediments more easily. As a result inverted trapezoid steel pipes are now in use at Sakurajima and Mt. Tokachi. The problem of screen blinding has to an extent been solved, simplifying the removal of deposited sediments. Although general concepts in debris flow breaker screen design have been gained from past research and testing but topics for further investigation still remain. Model tests indicate that sediments accumulate at screens angled similarly to the angle of repose. In reality however, the deposition gradient varies according to changes in sediment concentration in the freshet. Thus research is required into angle of repose.

4.1.3 Steel Cell Dams. The first steel cell dam commissioned in Japan was built in August 1991 on the Yotagiri River (length 16km, catchment area $42.7km^2$), a severely devastated mountain river in the upper reaches of the Tenryu River (see Photo.3).

Steel cell dams:

1. can be built within very short periods since the steel members can be readily assembled even in remote mountain areas with minimum labour
2. enable the use of local materials, reducing transportation costs and simplifying the work process
3. allow stones to be removed easily, making them suitable for areas subject to frequent debris flows

Photo 3 Yotagiri River steel cell dam

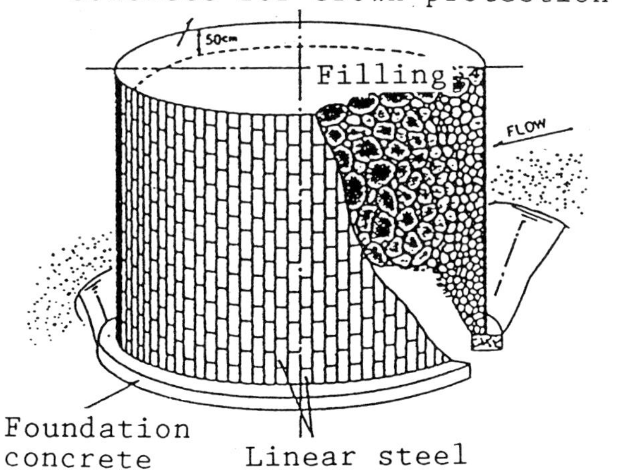

Fig. 5. Structure of steel dam

The steel cell is a steel sheeting cylinder piles filled with local materials and capped with concrete (see Fig. 5). Optimal effect is gained by staggering cells.

In terms of function steel cell dams are treated as being permeable type. For stability analysis calculations however they are considered similar to impermeable types due to the proportionally small opening. Since all cells function together as a whole, the fills are resistant to shearing deformation from horizontal forces.

Steel cell dams can:

1. capture flow front and reduce debris discharge
2. prolong the time taken from generation to reach alluvial fan
3. stop boulders and driftwood at flow front
4. covert debris flow into hyper-concentrated flow
5. reduce peak discharge of debris flow

An ITV camera is used to monitor the descent of debris flow and thus assess the performance of the cell dam. The upper reach face of each cell is also fitted with gauges to measure pressure, impact force and strain. As yet no debris flows have been observed so instead observation is being focused on sediment deposition behind the dam due to movements of sediments immediately above. There is slight grazing and other damage on the cell surfaces but no strain or other phenomena. Observations will continue hereon.

4.2 Slit Dams

Slit dams may have one or any number of slits. The purpose of these is to provide a sound sedimentation capacity. enabling slit dams to control sediments more efficiently than ordinary sabo dams in mountain rivers threatened by large sediment discharges during debris flows in debacles (see Photo 4).

Slit dams are used where sediment damage from small and medium freshets does not cause damage and a supply of sediments is needed in the lower reaches because of degradation. The slit is also suitable as a fish way.

Photo 4 Slit sabo dam (Fuji River)

Slit dams:

1. boost the capturing volume of a dam by allowing harmless non-flood sediments to descend rather than accumulate needlessly
2. enable sediments to continue to lower reaches in order to minimize river bed deformation during non-flood periods
3. temporarily accelerate deposition through the afflux action of slits during debacles and enable sediment discharge until the non-flood deposition level is restored by the latter half of the debacle and subsequent small and medium freshets
4. capture boulders at the flow front thus halting the flow itself.

Experimental research [7][8] has revealed that the volume of sediment controlled by a slit dam under 3) above increases as the slit itself narrows. A condition of slit height (h) is that a deposit shoulder should be formed by afflux from the dissipation of flow in the vicinity of the sabo dam. The volume can be determined using parameters describing the flow, such as Froude number, width at damsite, width at deposit shoulder, gradient of design sedimentation gradient, Manning coefficient of roughness and design discharge. This volume is equivalent to the volume of sediment between the non-flood deposition level and the deposition level for the design discharge shown in Figure 6 [7][8].

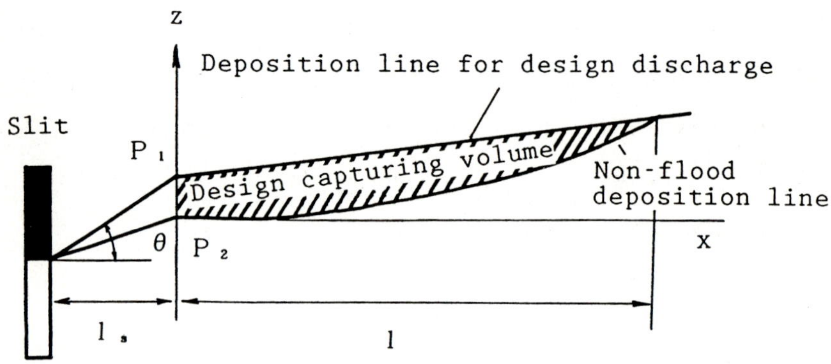

Fig. 6. Design capturing volume for slit dam [7]

The non-flood deposition line used in the plan and in the deposition line for design discharge are drawn at 1/2 of the gradient of the former riverbed from the upriver side of the slit base and the shoulder of deposits during design discharge respectively. Normally q $=35^o$ is used for slope gradient (q) downriver form the shoulder of deposits and treated as equal to the submerged angle of repose.

Design capturing volume may need to be restored artificially (by removing stones) if the natural process is inadequate. According to test results [7] blocking in a debris flow is highly likely if the slit width is narrower than 1.5 times the maximum gravel diameter. Partial blocking occurs when the slit width is more than two or three times the maximum gravel diameter, although peak discharge decreases.

In order to halt a debris flow with a degree of certainty (i.e. point 4) above) the slit width must be approximately less than or equal to the maximum diameter of the gravel in the flow. If the aim is to reduce peak discharge (point 3) above) then the slit should be at least two or three times the maximum gravel diameter. To have the depositing area empty during non-flood periods but capturing sediments during debacles the slit should be set to a width larger than 2 - 3 times the maximum gravel diameter permitting movement in medium or minor flows but smaller than 1.5 times the maximum gravel diameter permitting movement in a debacle [7].

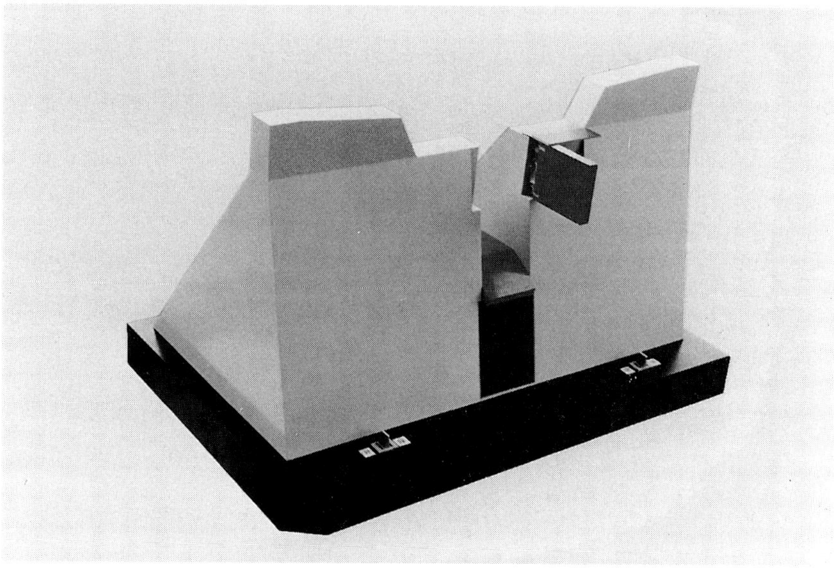

Photo 5 Model of sabo dam with gate

If a small slit is used in a slit dam whose purpose is point 3) above the hydrograph can be smoothed out to some extent but there is also the potential for abrupt discharge from deposited sediment, depending on the form of discharge. Research is currently underway into artificial sediment discharge control facilities (sabo dams with gates) where the slit has a gate so that sediments can be allowed through during non-flood periods but deposited during the June-July rainy season and the typhoon season [2] (see Photo 5). This should help solve the problem of abrupt discharge from deposited sediment and capture sediment during debacles by boosting capturing volume.

The mechanisms behind the generation, descent and deposition of "mysterious" debris flows, of which little was known previously, have been clarified considerably by recent research. This new-found knowledge has been widely used throughout Japan in the design of erosion control facilities for hazardous torrents prone to debris flows. Many of these facilities have proved successful in preventing disasters.

Nevertheless there remain a number of areas requiring further research. Even now, erosion control facilities are occasionally partly destroyed by debris flows. Research should continue into ways to determine the volume of flows and estimate the external forces on sabo dams in order to enhance dam design.

The new permeable sabo dams (e.g. steel slit dams) have come into widespread use, their performance established through experiments. Yet these dams have only a short history of actual use. Their performance has not been verified in the field. Permeable sabo dams are also capable of controlling driftwood discharge, a major problem in debris flow disasters. Future field verification is most important. At the same time we must tackle problems of planning, design, construction and maintenance.

REFERENCES

[1] Japan Sabo Association: Sediment Disaster Countermeasures, 1989.
[2] The Japan Society of Erosion Control Engineering Eds: Lectures on Erosion Control Science. Vol. 6-1 Countermeasures for sediment-related disasters alluvial fans, debris flows and others (1), Sankaido Co., pp 106-107, 1992.
[3] Sediment Disasters Annual Report Committee and Sabo Technical Centre (Foundation): Sediment Disaster Facts (1983-1992).
[4] Sediment Control Division, Sediment Control Department. Ministry of Construction: Technical Standard for the Measures against Debris Flow (draft), 1990.
[5] Masayuki Watanabe, Takahisa Mizuyama et al: Experiments on sabo structures for debris flow, Civil Engineering Journal, 22-2, 1980.
[6] Masayuki Watanabe, Takahisa Mizuyama and Shinji Uehara: Use of Erosion Control Facilities as Debris Flow Countermeasures, Journal of the Japan Society of Erosion Control Engineering No. 115, 1980.
[7] Hiroshi Ikeya and Shinji Uehara: Experimental Study about the Sediment Control of Slit Sabo Dams. Journal of the Japan Society of Erosion Control Engineering No. 114, 1980.
[8] Takahisa Mizuyama and Sohei Abe: A Study on Sediment with a Slit Sabo Dams. Technical Memorandum of PWRI No. 2851, 1990.
[9] Yoshiharu Ishikawa, Michiya Irasawa and Akihisa Fukumoto: The study of the Effect and Operation of Sabo dams with an Artificially Operated Gates. Technical Memorandum of PWRI No. 2943, 1991.

DISCUSSION

Julien: I am curious about the steel cell dams.

1. Aren't they more vulnerable to boulder impact than slit Sabo dams? 2. Do you expect cohesion problems due to volcanic ashes?

Jaeggi: 1. Can debris flow screens be made very long to ensure that successive pulses are stopped? 2. Does management of flexible (permeable) structures imply the full excavation of debris flow deposition or is a clearing of the screen sufficient?

Yamada: 1. It was not necessary in the presented cases. Cost will be high with greater length. Basically a great length would lead to higher safety. 2. For safety reasons full excavation is common. Clearing of screens and evacuation of boulders is essential.

Torrent Check Dams as a Control Measure for Debris Flows

Martin N.R. Jaeggi and Stefano Pellandini

Laboratory of Hydraulics
Hydrology and Glaciology
Federal Institute of Technology
Zurich, Switzerland

1. Introduction

People living in mountain areas have been threatened by natural hazards like avalanches, floods and debris flow for centuries. A popular way to protect settlements against floods and debris flows is to build torrent check dams. Early structures are reported to have been built in the European Alps in the 14th century (see Bundesministerium, 1984). It is difficult or almost impossible to find out who started this type of construction and what the design policy was which then was applied.

In the 19th century, building torrent check dams triggered common in the Austrian and Swiss Alps. It seems that from these countries this torrent control concept has been brought to other countries with similar problems. According to H. Ikeya the German engineer Drijke (1873) and the Austrian engineer A. Hofmann (1905) introduced it in Japan.

The construction of railway lines at the end of the 19th century across the Alps triggered a lot of control works in torrents. During the recent years, increased land use in mountain regions raised the need for a better protection. In densely populated areas like in the European Alps or Japan many series of check dams have been built.

The design of these structures relies mostly on observation and professional experience. It is to a large extent intuitive. Different approaches to the design problem can however be distinguished. They are presented hereafter and the effect of structures conceived according to the different concepts on debris flow dynamics or debris flow prevention is discussed.

2. The River Engineering Concept

In heavily trained rivers of a slope higher than about 0.3% often drop structures are placed at regular intervals. At their location, they fix the longitudinal profile of the river. In between, the slope becomes usually flatter than the valley slope. The drop structure has to cope with the corresponding difference in energy levels.

Although check dams were probably first applied in torrents and then in rivers, the design concept in torrents is quite often derived from the river engineering concept.

A well reported early experience in terms of a systematic regulation of a torrent using such a method is the training of the Rio Lis near Leira in Portugal (de Campos Andrada E., 1980). Many disastrous floods were reported (1475, 1596, 1600, 1617, 1646). In 1840 a systematic concept was attempt to the landowners. However, aggradation of the bed was the consequence. It is reported that only headworks along the river and reforestation reduced the sediment input and thus the schema was more successful.

2.1 Valley Incision

In a narrow V-shaped valley which is typical for eroding reaches in steep mountain rivers any lowering of the bed level will induce landslides on the side slopes. If the slopes correspond to the natural angle of repose up to the boundary, then a parallel erosion occurs on the side slopes. In order to stabilize such reaches, it is necessary to fix the valley floor on a certain level (fig. 1).

Different engineers formulated the possibilities to stop such incision processes by building transverses structures during the 19th century (Hofmann, 1913; de Preux, 1918).

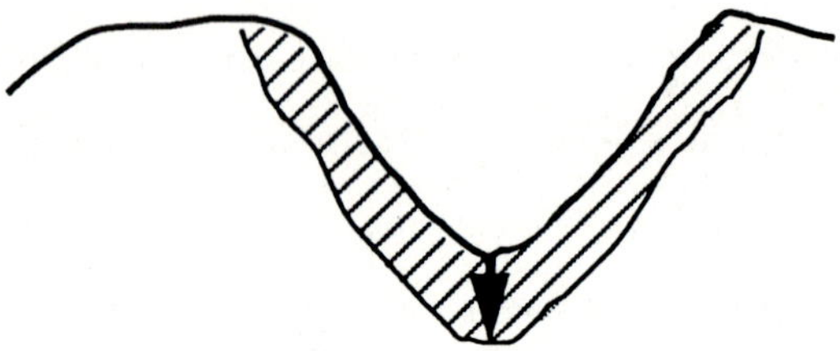

Fig. 1. Effect of bed degradation in a V-shaped valley on slope erosion

2.2 Fixed Points and Dynamic Reaches

Fixation of the valley floor on steep slopes is usually realized by building check dams at regular intervals. The check dams fix the longitudinal profile of the river or torrent locally. In between, the channel is still self formed. Since normally the check dams are raised somewhat above the original bed, the sediment supply from the bed and the side slopes is reduced. In the dynamic reaches between the check dams a local slope flatter than the valley slope therefore develops (fig. 2).

The first example of a gully stabilization with a check dam series is reported to have been introduced near Brixen. It was built between 1650 and 1689 (see Bundesministerium, 1984). Others examples of check dams that were meant to stabilize the bed of a valley or a gully in this way were built in Tirol (Austria) just after the year 1800. Systematic application was not yet widespread until the first half of the 19th century. The dams were usually low and often made of wood. A guide line for the construction is due to Duile (Austria, 1826).

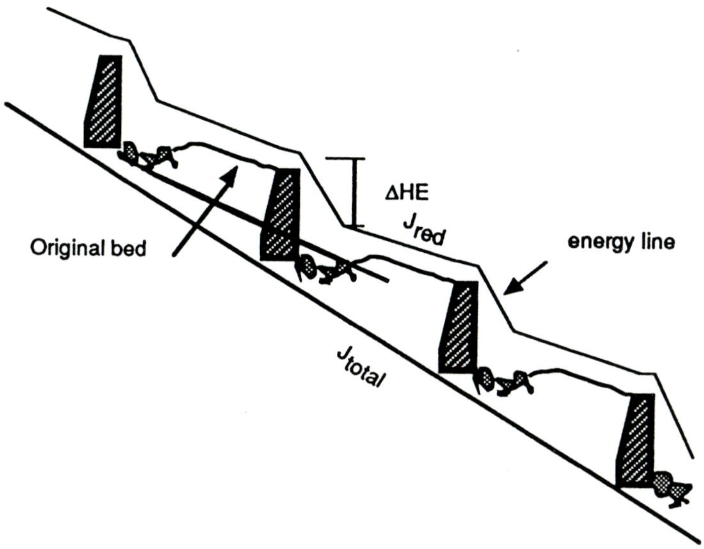

Fig. 2. Schematic representation of a check dam series built according to a river engineering concept. In the dynamic reaches between the check dams a flatter slope than the valley slope develops, mainly by deposition. The scour problem at the toe of the dams requires additional control measures

2.3 Sediment Transport Implication

Since the main goal of such a regulation scheme is to prevent a further lowering of the valley floor and thus to reduce erosion rates, the sediment rates which will be predominant in the formation of the new channel will be substantially smaller than in the original eroding valley. If there is no supply from upstream and the control of the side slope has become totally effective, even a zero supply must be taken into account. A quite low reduced slope between the steps may be the consequence.

As a function of the difference between valley and channel slope and the spacing of steps at each check dam a certain fall height results.

2.4 The Scour Problem

For more or less clear water flow, the energy head corresponding to the fall height has to be dissipated, which results in a scour hole of a certain extent. Basically, the foundation of the check dam should be as deep as the scour depth for a design discharge. This condition is almost impossible to fulfil if a substantial excavation in the unstable side slopes has to be made. Quite often, the foundations are therefore insufficient. Often, scour depth is reduced by adding coarse boulders on the bed. Narrow spacing and general raising of the new bed may by-pass the excavation problem , but result in very costly solutions.

Aprons at the toe of the dams may divert the flow and prevent scour. Stress on such aprons is however high. If they are made from loose blocks, a scour may still form at very high flows. If the scour is prevented, energy dissipation occurs on the following reach. For check dam series with rather narrow spacing, each dam may have an upstream effect like a sill in a stilling basin (Δz in fig. 3).

If the energy dissipation at the toe of a check dam is totally prevented by the insertion of such aprons, then the total slope is still dominant for the erosive capacity of the flow on the movable bed between the check cams. In such a case, the increase of resistance on the bed has to be almost the same as if the check dams would not be there. This extreme situation however does not really occur, because the jet deflection on the apron always causes some energy dissipation.

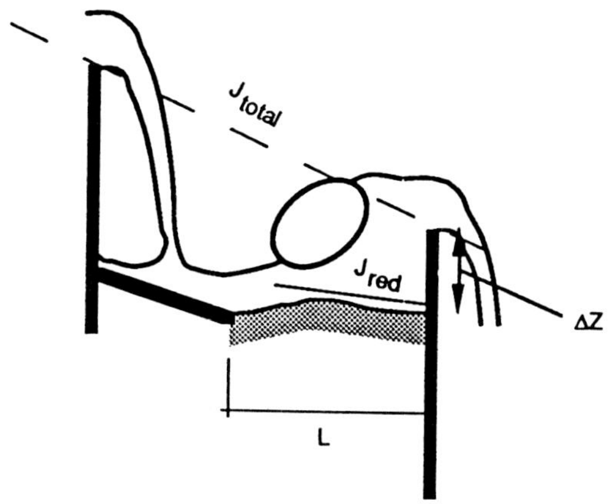

Fig. 3. Effect of an apron at the toe of a check dam in a narrow spaced check dam series. The space between the check dams becomes a stilling basin

2.5 Debris Flow Guidance

Where a steep mountain stream or torrents has been regulated by the insertion of check dam series, the main objective may have been flow and especially debris flow guidance. In such a situation, not only vertical instability but also lateral instability may be a problem. The wing walls of check dams take over the function of groynes in a river or concrete lining of debris flow canals. Since normally the spacing of check dams on steep slopes is low, the check dams alone may be sufficient to prevent lateral instability.

Sometimes, debris flow guidance may be the major purpose of control works. It is known that a debris flow forms levees where its flow section is not defined. They simply form because water drains laterally, and subsequently these parts cannot move any more. Check dams may have the purpose to define where these levees should form. Once they are present, they will guide further debris flows.

If debris flows are still to be expected from upstream, special attention has to be paid to the conception of check dams. The flow section has to be more rounded and sharp corners should be avoided.

3. The Sediment Detention Concept

Where debris flow threaten to cause major damage in the main valley, a sediment detention concept is sometimes applied. Large dams create considerable detention volumes.

Sediment detention dams were probably the first attempts to control torrents in the Alps. Allegedly the first check dam was built in Tirol 1537 (Stacul P., 1979). The oldest still existing structure dates back to 1612. Originally it was 20 m high and has been over the years successively raised to an actual height of about 35 m with a width between 4.5 and 10 m. The structures were made of dry masonry, usually placed at the downstream and of flatter reaches in order to maximize the detention volume. Documentation on simple studies dates back to 17th century in Trentino (Italy, see Mariani, 1686). A variant of this principle is the idea to allow potential erosion of the stored volume by minor flood using slot dams or similar structures. Attempts (see Leys E., 1976) were made by Venetz (Switzerland, 1825), Gras (France, 1857) and Demontzey (France, end of 19th century).

Quite often, it can be postulated that detention volumes are sufficient to retain the yield carried by the torrents over years or decades. Where this is not the case, the detention volume has to be periodically excavated.

Where excavation is not possible, for instance because access is impossible, after some time a situation may be reached which corresponds to the one before the construction of the dam. The effect of the dam was then only a temporary one, although lasting over 50 or 100 years. However, the situation may be worse now. The deposition behind the dam has become a new potential danger should the dam break.

It is sometimes claimed that in this situation a second series of check dams has to be built on top of the first one. As for the first series new detention volumes can thus be created. If important constructive means are involved, this procedure may appear to be successful. However, it means that the products of weathering are kept on top of the slopes. Again, the potentials of loose materials are increased and damage may be worse in case of an extreme event endangering the structures.

4. The Supporting Wall Concept

A hillslope consisting of loose material cannot be steeper than corresponding to the natural angle of repose. Where land use does not allow to respect this conditions, supporting walls are built. A series of regularly spaced supporting walls allow to stabilize the hillslope at a steeper angle (fig. 4)

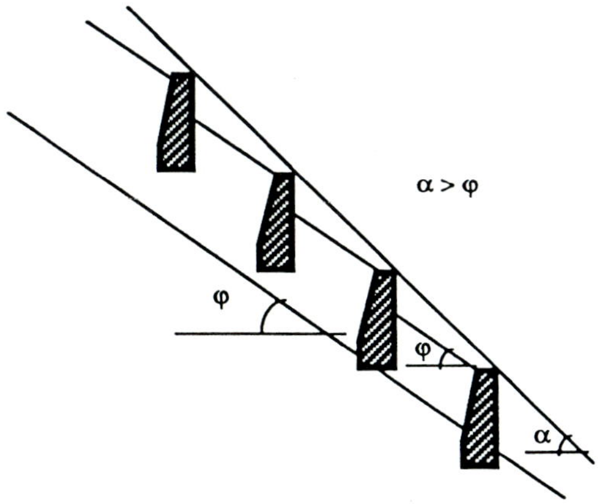

Fig. 4. Stabilization of a slope at an angle steeper than the original one

More or less intuitively, this concept has been applied to scree slopes. Where weathered material is supplied from rock slopes situated above, the material is continuously detained by such structures and the slope steepened according to fig. 4). Check dams which have been inserted in debris flow gullies which had formed in such scree slopes may have been built according to such a supporting wall concept.

4.1 Limited and Unlimited Sediment Supply

If such a construction scheme can be extended up to the upper edge of a catchment which supplies sediment, then one may talk about a limited supply situation. A gully which starts to develop may be virtually closed by check dams. This procedure may be particularly appropriate if the erosion starts in older alluvial deposits. Fixation of the torrent bed and eventually fixation of the slopes will prevent any further erosion. The supply of sediment is limited.

In contrast to this, in areas with recent deposits sediment supply may often to be considered as unlimited. Weathering of bare rock is a more or less a continuous process. Scree slopes at the bottom will be fed continuously. Stabilizing the scree slopes and/or the gully draining the catchment by supporting walls will only have a temporary effect, although lasting over some decades. Sometimes, even if the gully had formed in a rocky environment, check dam series have been built in such situations. It was hoped that debris flows fed by the material of the scree slopes would be prevented by these dams.

4.2 Debris Flow Dynamics on Unsaturated Steep Slopes

A quite common feature on steep mountain slopes is debris flow initiation on scree slopes presenting almost the natural angle of repose. According to Takahashi (1991), in such conditions the material can never be saturated with water, otherwise landslides would have occurred. The stability for saturated material corresponds to a far flatter angle.

Frequently, these debris flows are triggered by the inflow of water which is gathered by a small funnel like catchment in the rocky portion above the scree slope during heavy precipitations (see fig. 5). Only small discharges can completely infiltrate at the interface between rock and loose material. The infiltration rate along the debris flow channel will determine which part of the inflow will be surface flow. If the material of the scree slope is homogeneous and the moisture content the same, then the infiltration rate is the same along the channel and corresponds to the permeability of the material for unsaturated conditions. If no effect of boundary conditions has to be considered, the hydraulic gradient for vertical infiltration can be taken as equal to 1. The surface flow will therefore decrease along the channel and disappear after an infiltration length L. This is analogous to what happens for gravity irrigation.

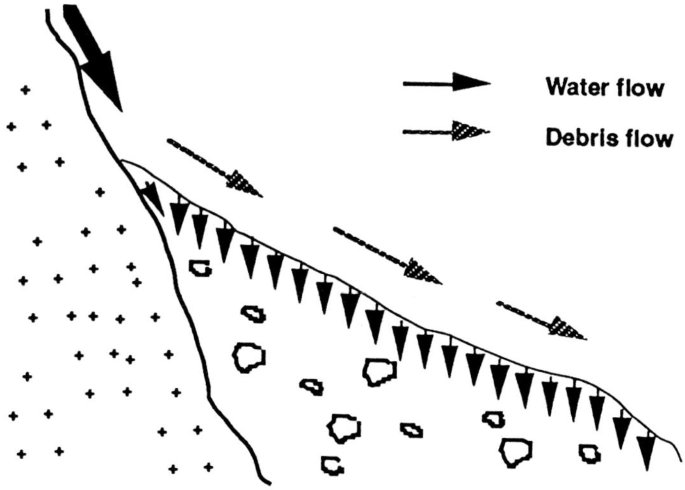

Fig. 5. Triggering of debris flow near the interface of rock and an unsaturated scree slope

On steep slopes, surface flow will immediately pick up sediment. Soon, the maximum possible volume concentration of about 0.63 (Takahashi, 1991), will be reached. At a slope corresponding to the natural angle of repose, it may be expected that this may have happened after a very short distance. A debris flow is therefore triggered by surface flow right close to the rock and scree slope interface.

Along the channel, the debris flow looses water because of infiltration into the scree slope. Since it had already reached the maximum volume concentration it must loose sediment as well. At least part of the deposition will be in the form of levees, since deposition is also favoured by lateral exfiltration of water out of the debris flow body. Erosion of the channel is therefore limited to the top part where the debris flow develops from surface flow.

In such a situation it is therefore useless to insert check dams in the gully in order to prevent gully incision. The concept in fig. 1 does not apply here.

4.3 Debris Flows Developing at the Toe of Scree Slopes

Situations may occur where all the surface runoff from small rocky catchments may totally disappear in the scree slopes even at extreme events because permeability is high and the rocky catchment small. Surface flow occurs then only at the toe of the scree slopes, especially if they lie on rock. Already the subsurface flow in the scree slope and the flow reappearing at the surface may have gathered in a steep gully.

If the gully itself is unerodable, then the flow can only pick up material from the scree slope. In the region of the spring the material will be fully saturated and therefore start to slide into the gully. The resulting mixture of water and solids will again correspond to the maximum volume concentration of 0.63. A debris flow therefore starts. In the rocky gully, without any further supply of water, it will just move downstream, without any further increase or decrease in size, except for an eventual development of pulses similar to roll waves (Davies, this volume).

Under more or less steady conditions, the supply of water may be again to be considered to be the limiting factor on debris flow growth. The conditions at the toe of the scree slope will allow the water appearing at the surface to pick up immediately enough material corresponding to saturation and a mature debris flow to form (see fig. 6). Building check dams at the toe of the scree slope and/or in the gully will again just have a temporary effect. Small detention volumes are then created which have first to be filled until the antecedent slope conditions are re-established. After that, debris flows still develop and move downstream as if the dams would not exist.

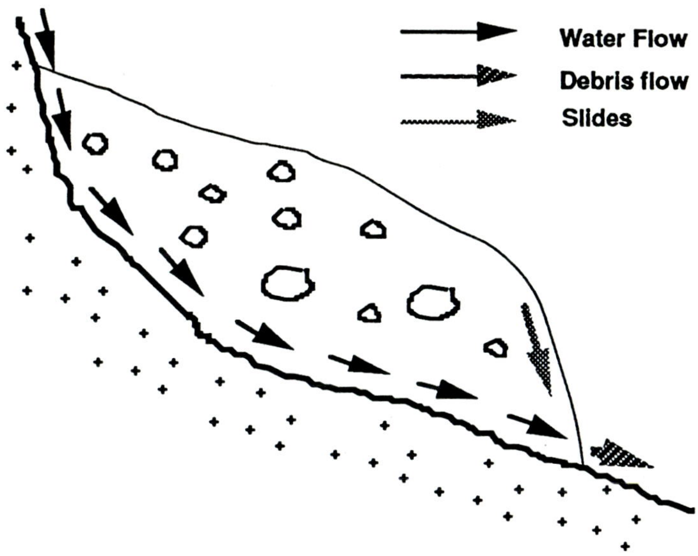

Fig. 6. Debris flow formation at the toe of a scree slope by subsurface flow

5. Prevention of Debris Flow Amplification

If a debris flow is mainly formed by a release of a stored volume of a water and sediment mixture then the instantaneous peak discharge may by far exceed the one corresponding to processes described just before, for comparable sizes of catchments obviously. Landslide fluidization, breakage of a temporary dam and bed fluidization are the three processes which are dominant in development of large debris flows and which can all be considered as such a release of stored water and sediment mixtures. On to the first two processes the construction of check dams will have little effect to prevent the phenomenon to occur, since the origin of the process is outside of the channel or gully. Only the process of bed fluidization will therefore be considered in this chapter, since check dams may be quite effective to prevent this phenomenon to occur or at least to reduce the effects (Jaeggi, 1993).

5.1 Debris Flow Amplification

In a certain range of steeper slopes, a bed layer of a substantial depth may be subject of fluidization by a small debris flow or even clear surface water flow (see fig. 7). If such a small surface- or small debris flow resulting from a short period with intensive rainfall in the upper catchment triggers the release of such a potential, then one may talk about debris flow amplification. In small but steep catchments, the amplification resulting from such a bed fluidization may be considered as particularly important.

Takahashi's (1991) eq. 3.1.10 allows to determine the amplification of flow depth from bed fluidization. Fig. 8 shows an application for clear water input and a stony bed. For a slope less than about 15%, amplification is negligible. Between slopes corresponding to 40 and 45%, the amplification increases asymptotically. Above a limiting value of about 45% no more significant

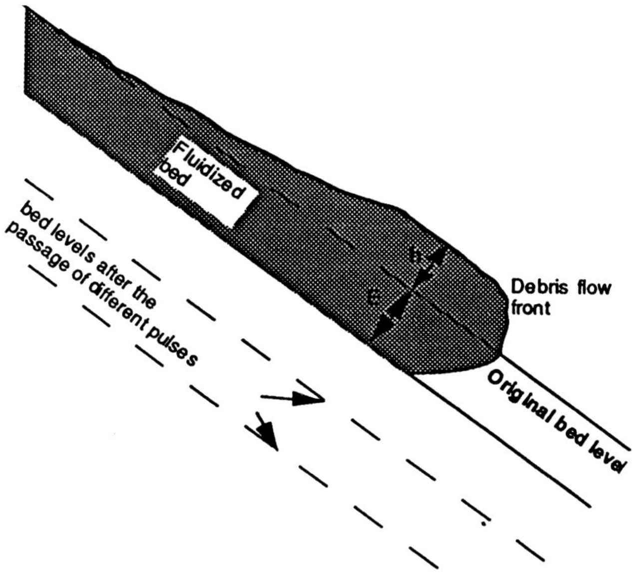

Fig. 7. Schematic illustration of bed fluidization by a debris flow (E= erosion/fluidization depth, h= flow depth) Successive erosion by following pulses

values can be found with this formula. This limiting value corresponds to the minimum condition for the occurrence of a landslide of a saturated mass of loose material.

The addition of water and sediment from the bed will increase the debris flow depth. In a downstream direction, the amplification according to fig. 8 should therefore continuously increase, specially since the mixing of water flow with the solids will have a retarding effect and thus causes further increase in flow depth. However, a maximum volume concentration of about 0.63 prevents an indefinite growth of a debris flow. The bed usually has a higher concentration of solids. It can be fluidized only if there is a supply of excess water. Because of the compensating effect of increased flow depth and limiting concentration, it is understandable why often it is reported that bed

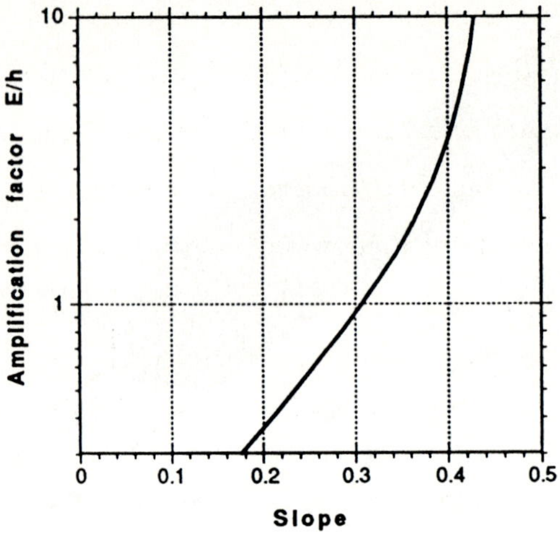

Fig. 8. Ratio of bed fluidization depth to surface flow depth, acc. to Takahashi (1991), eq. 3.1.10

erosion by a debris flow (which rather should be regarded as bed fluidization) is constant along a certain torrent channel, as illustrated in fig. 7.

An extensive study (Haeberli et al., 1991) on debris flows which had occurred in Switzerland in 1987 allowed to explain the main aspects of the origin of these events and of the Val Varuna event near Poschiavo in particular. In 1834 a similar event had already occurred. Later, it was attempted to stabilize the channel of the torrent by masonry check dams. It seems that after some years they prevented minor events to occur. In 1987 a long rainfall period however led to bed saturation. On July 18/19, several rainfall intensity peaks must have triggered several debris flow pulses. The total amount of erosion depth (or fluidization depth) was about 5 to 10 m. This fluidization depth was reasonably constant along the channel. The volume of sediment deposited on the fan was approximately 350.000 m^3. About 15% of this volume corresponded to the backfilling of the old dams, all destroyed and entrained by the debris flows. In contrary to the first assumptions the flanks did contribute only little to this mass. Obviously, they had however become unstable and the supply of sediment to the channel from the flanks started again immediately after the event. One single debris flow pulse did not entrain more than about $50.000\,m^3$, despite the fact that the bed was fully saturated. This supports the idea that the supply of clear water resulting from high instantaneous rainfall was the limiting factor.

5.2 The Effect of Check Dams

A particular effect of torrent check dams is to produce a reduced bed slope between the structures compared to the valley slope, which is then predominant at least for some years after the construction. Referring to fig. 8, the reduction of slope may result in an important reduction of the amplification factor. Thus bed fluidization is reduced or even prevented. If a series of check dams are installed in a steep torrent, sediment produced in the flanks and moved by smaller floods will backfill the dams, what will result in the described slope reduction. Therefore, a clear water flood or a debris flow, which in the uncontrolled torrent would have produced substantial bed fluidization and thus resulted in a heavy debris flow event, may not be able now to fluidize the bed because for the reduced slope the amplification factor is small.

Only if the backfilling of the check dams is such that more or less the original valley slope is reached again by the deposition, then a partial fluidization of these deposits seems possible again. Such an evolution is possible only if intermediate floods, which normally would continuously evacuate the supplied sediment, are absent. In case of the quoted Val Varuna events, a particularly large water storage capacity of scree slopes seems to reduce the occurrence of floods. Only extremely long precipitation periods can produce substantial surface flows. In such conditions the bed of the main channel will then be saturated anyway.

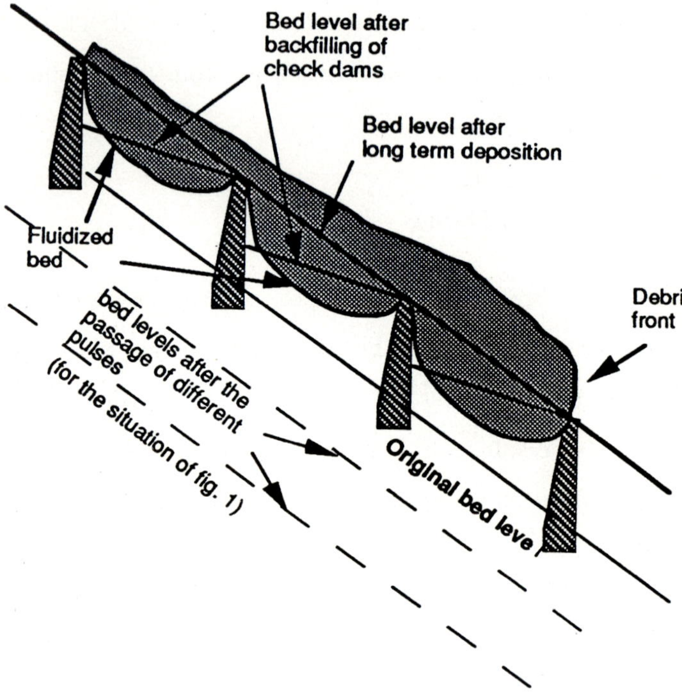

Fig. 9. Schematic view of a debris flow action on a torrent bed secured by check dams. Bed fluidization is possible only if before the event the slope between the dams has reached a value similar to the valley slope. Normally, this slope value is reduced and bed fluidization not possible. This picture therefore illustrates an extreme situation. Compared to fig. 7, channel incision is prevented anyway; bed fluidization by following pulses is no more possible

Massive concrete check dams are expected to have a stabilizing effect even in such a case. By analogy to the effect of a series of regularly spaced steps in a mountain stream (see Whittaker, 1982), the boundary between stable and mobilized bed may have a shape as shown in fig. 9. For the first debris flow pulse passing over the the saturated deposition, fluidization will occur, except for the regions close to the dams, down to a depth corresponding to the amplification of fig. 7. After that, the bed will be rearranged by the flow in order to present a reduced slope again, and a further fluidization is not possible any more. This is obviously true only if the concrete structures are able to support such dynamic stresses.

If regular floods or small debris flows occur while the slope is still reduced, then obviously any bed fluidization is prevented. In torrents where bed fluidization is the reason why comparatively small debris flow are amplified into a disastrous event, the construction of a check dam series is therefore a suitable counter measure.

In the Val Varuna, soon after the event of July 18/19, 1987, the construction of about 50 check dams, distributed on a length of 1.5 km, was started. The design relied mostly on experience and was somewhat intuitive. However, a careful structural design resulted in quite massive reinforced concrete structures. Backfilling with sediment, mainly issue form the unstable flanks, was immediate. However, compared to the valley slope of about 37%, a reduced slope of now about 22% was established. According to fig. 8, if a flow depth of a triggering flood of about 1 m is assumed, bed fluidization is then insignificant.

Thinking back to the history of the Val Varuna between 1834 and 1987, it must be kept in mind that a slow but continuous deposition of sediment between the check dams is quite probable. Supply from the flanks is still abundant. The occurrence of a single debris flow pulse amplified by bed fluidization is therefore possible in a couple of years and decades. Often, after such periods, people have almost forgot these events.

In contrary to what happened in 1987, the triggering of more than one pulse seems to be impossible now with the concrete dams. Fig. 9 illustrates schematically that the erosion levels of fig. 7 can not be reached any more.

Even the collapse of one or several dams would not increase substantially the volume of sediment involved. This reasoning is based on the assumption that the triggering rainfall intensities will not be higher than in 1987. A more detailed risk analysis would have to concentrate on this point.

6. Conclusions

The design of torrent check dams is mainly a structural one. The concept applied in terms of erosion and debris flow control is mainly an intuitive one. However, valley incision in older deposits of loose material may be prevented by applying traditional river engineering concepts. Debris flow amplification may be prevented by using recent results from research on debris flow. In other situations, and especially on very steep slopes and unsaturated scree slopes, the use of check dams does not seem very appropriate.

REFERENCES

Bundesministerium für Land- und Forstwirtschaft, (1984) 100 Jahre Wildbachverbauung in Oesterreich, 1884 - 1984

Davies T.R., (1993) Large and small debris flows - occurence and behaviour, Departement of Natural Resources Engineering Lincoln University, New Zealand, Proceedings International Workshop on Debris Flow, Kagoshima

de Campos Andrada E., (1982) 80 anos de actividade na correcÇáo torrencial - hidraulica florestal - (1901-1980), Ministério da agricultura, comércio e pescas, DirecÇáo general das florestas, Lisboa

de Preux H., (1918) Etude pratique sur la construction des routes de montagne et la correction des torrents dans les régions élevées, Paris, Neuchâtel

Haeberli W., Rickenmann D., Zimmermann, M. und Rösli, U. (1991) Murgänge. Ursachenanalyse der Hochwasser 1987, Ergebnisse der Untersuchungen. Mitteilung des Bundesamtes für Wasserwirtschaft Nr. 4, Mitteilung der Landeshydrologie und -geologie Nr. 14. Eidg. Druck- und Materialzentrale, Bern.

Hofmann A., (1913) Aus den Waldungen des fernen Osten, Wien und Leipzig

Ikeya H., (1979) Introduction to Sabo works, The Japan Sabo association, Tokio

Takahashi T., (1991) Debris flow, International Association of Hydraulics Research Manual, Balkema, Rotterdam/Brookfield

Jaeggi, M., (1993) Torrent Check Dams and Debris Flow Amplification, Proceedings of the IAHR conference, Tokyo, 30 August - 3 September

Killian H., (1990) Dokumente und Materialien zur Geschichte der Wilbach- und Lawinenverbauung in Oesterreich, Mitteilungen der forstlichen Bundesversuchsanstalt Wien, Wien

Leys E., (1976) Die technischen und wirtschaftlichen Grundlagen in der Wildbachverbauung der grossdoligen und der kronenoffenen Bauweise, Dissertationen der Universität für Bodenkultur in Wien, Wien

Mariani M., (1686) Trento con il Sacro Concilio ecc. Augusta

Schnitter N., (1991) Die Geschichte des Wasserbaus in der Schweiz, Olynthus, Oberbzberg

Stacul P., (1979) Wildbachverbauung in Südtirol gestern und heute, Sonderbetrieb für Bodenschutz, Wildbach- und Lawinenverbauung, Autonome Provinz Bozen Südtirol, Bozen

Tomé G., (1937) Enquête international sur la correction des torrents et sur la restauration des montagnes en Europe, Institut international d'agriculture, Roma

Whittaker, J. (1982) Flow and Sediment Movement in Stepped Channels. A thesis submitted in partial fulfilment of the requirements for the degree of Doctor of Philosophy, Lincoln College, New Zealand

DISCUSSION

Davies: 1. The abscissa of Fig. 8 is "slope": is this definitely the local bed slope between steps or could it be the average energy line slope? If it is the latter, the application cannot be prevented by check-dams (unless they have scour aprons)?

2. The 1834 and 1987 storms at Varuna were similar. In 1834, $50,000\,m^3$ of sediment yielded; in 1987, $350,000\,m^3$ yielded. The 1987 event restored the long-term average sediment yield from the catchment, that had been suppressed by the check-dams. Equally, rebuilding the check dams will have the same effect again?

Jaeggi: 1. The assumption that the reduced slope between dams is determining the amplification factor may be considered as a first approach. In a next step, it may be verified in specific model tests. 2. In the Val Varuna, floods triggering debris flows are rare. Without check dams, each pulse will find abundance of material to be fluidized. The volume transported downstream will depend on the triggering discharge and the amplification factor. Between 1920 and 1987, the old check dams did indeed retain material and increase the potential. This volume was about only 10–15% of the total volume of $350,000\,m^3$ deposited on the fan in 1982.

In fact, during the 1987 event, the existing dry masonry check dams could not resist the debris flow. The dams themselves and the material stored behind them became part of the debris flow. However, altogether this volume contributed only to about 10 to 15% of the total mass moved by different debris flow pulses. After the completion of the new check dams, made out of reinforced concrete, the situation may look differently. If the extreme situation ever is re-established, that the depositional slope between the steps is the same as the original slope, then a major debris flow event is still possible. It is however hardly conceivable that these massive structures would be immediately liquefied and loose their function immediately. As soon as the amplified debris flow is able to move away substantial volumes of the deposited material, the slope between the steps will reduce again and the phenomenon will be slowed down.

If, in the worst case, some of the dams fail, then the situation will be locally equal, but not worse, than in a natural situation, since the driving parameters are about the same.

Julien: On Fig. 3, an apron is recommended to prevent toe scour downstream of check dams. How thick should they be when boulders overtop check dams and impact the apron?

Jaeggi: The apron sketched in fig. 3 is normally applied in streams carrying little sediment only. In case of a torrent carrying a lot of sediment, this system is basically conceivable only if the series of check dams extents over the whole erosional reach and the check dam series prevents further massive erosion.

On the Dynamic Impact of Debris Flows

Aronne Armanini

Department of Civil and Environmental Engineering
Faculty of Engineering
University of Trent - Italy

1. Introduction

The literature concerning dynamics impact of debris flows is lacking, although the problem is of definite interest. Since protection structures, like debris flow barriers, debris racks and fences and debris breakers, must be designed to withstand dynamic impacts.

Two different approaches have been proposed in the technical literature. According to common practice in fact, the dynamic force exerted by a debris flow on a fixed structure is assumed proportional to the hydrostatic pressure on the structure: experimental evidences on collapse of some check dams structures seems to support this approach [Lichtenhahn, 1973, Aulitzky, 1990]. From a theoretical point of view this procedure is difficult to be sustained, the over-pressure being in fact related to hydrodynamic actions rather than to hydrostatic pressures.

According to another approach the dynamic impact is assumed proportional to the square of flow velocity [Mizuyama and Ishikawa, 1988]. Also this approach presents some uncertain aspects, because it does not account for unsteadiness of the phenomenon.

In order to clarify the problem, an experimental investigation was carried out by Armanini and Scotton, [1992], and a more rationale approach was proposed. A debris flow was simulated in a tilted flume at the end of which a vertical gate was placed. The dynamic impact was measured by a pressure transducer. The debris was generated by the release of a fixed volume of a mixture of water and plastic materials of different characteristics and at different concentration.

Basically two types of impact have been observed: according to first type, at the time of the impact the flow is completely deviated so as to form a vertical jet-like bulge; according to a second type of impact, at the time of the impact a reflected bore is formed and propagated upstream.

Models of the behaviour of the mixture and valuation of dynamic impact forces are given trough the analysis of momentum balance in both situations.

2. Theoretical Analysis of Dynamic Impact

The most interesting and severe case of dynamic impact is that of the impact of a surge over a rigid wall. In order to investigate the phenomenon it is necessary to clarify the dynamics of propagation of a sharp front. This is valid both in case of real debris flow over a granular bed as well as for a debris current moving over a fixed bed.

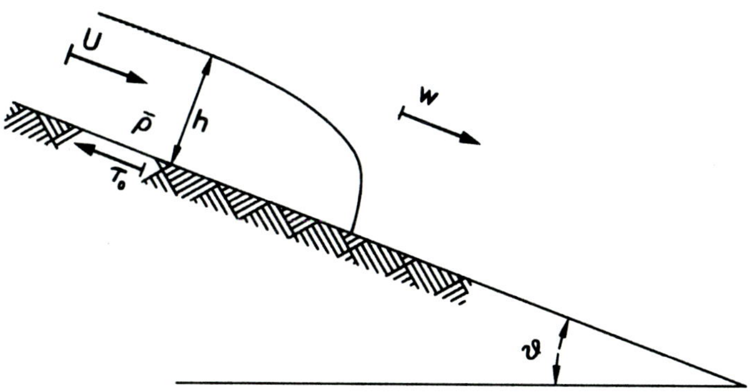

Fig. 1. One dimensional scheme for debris-flow and notations

The phenomenon can be theoretically analysed under the hypothesis of one-dimensional (depth integrated) flow, considering the mixture as a homogeneous fluid. In this hypothesis, the motion of the mixture is described by global balance of momentum and mass:

$$\frac{\partial U}{\partial t} + U\frac{\partial U}{\partial x} + g\cos\theta\frac{\partial h}{\partial x} = g\sin\theta - \frac{\tau_0}{\bar{\rho}\,h} \tag{1}$$

$$\frac{\partial h}{\partial t} + h\frac{\partial U}{\partial x} + U\frac{\partial h}{\partial x} = 0 \tag{2}$$

where $\bar{\rho}$ is the mean density of the mixture, U is the depth averaged velocity, h is the flow depth, θ is bed slope angle, and τ_0 is the bed shear stress. τ_0 can be related to mean velocity U, trough a uniform flow formula.

The general solution of set 1 - 2) is not known. However two important problems can be analitically trated: *dam break for an ideal fluid* and *steady front of real fluid propagating with uniform celerity* .

problem of dam break over horizontal bed for an ideal fluid.

Fig. 2. Scheme of dam break problem, for dry downstream riverbed and notations

For this purpose eqs.1) and 2) can be rewritten considering the speed of a small gravitational perturbation:

$$c = \sqrt{gh\cos\theta} \tag{3}$$

After derivation of eq.3) with respect to time and space, and after substitution in eqs.1) and 2), one obtains:

$$\left(\frac{\partial}{\partial t}+(U+c)\frac{\partial}{\partial x}\right)(U+2c)=g\sin\theta-\frac{\tau_0}{\bar{\rho}} \tag{4}$$

$$\left(\frac{\partial}{\partial t}+(U-c)\frac{\partial}{\partial x}\right)(U-2c)=g\sin\theta-\frac{\tau_0}{\bar{\rho}} \tag{5}$$

If the second term of eqs. 4) and 5) is set equal to zero, that is in the case of ideal fluid over horizontal bed or assuming local balance between flow resistance and longitudinal component of the weight, an analytical solution of the equations is possible and it allows some physically significant comments on the phenomenon under investigation.

In particular the phenomenon of the instantaneous abatement of the upstream gate, under the hypothesis of dry downstream riverbed, is solved analytically yielding:

$$w=2c_0=2\sqrt{gh_0\,\cos\theta} \tag{6}$$

where w is the celerity of the front and h_0 is the undisturbed flow depth upstream the gate.

In case of real fluid, and particularly in case of debris-flow, one must expect that the relationship between front velocity and initial flow head diverge from the theoretical values of eq. 6).

In the following Fig. 3, experimental values of front celerity (dimensionless with respect to $\sqrt{gh_0}$) are reported as a function of initial flow depth, in case of a dam break for different highly concentrated mixtures of PVC and anionic resin particles [Armanini and Scotton, 1992].

Depending on the ratio between gravity forces and resistance forces, the front celerity might be smaller or higher than the value predicted by ideal case (eq.6).

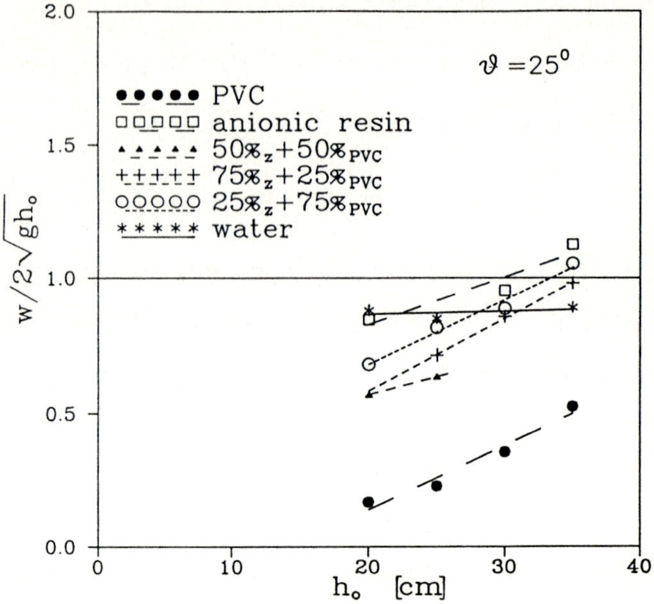

Fig. 3. Values of the dimensionless front celerity versus undisturbed flow depth h_0, for a mixture of granular material of different composition; $\theta = 25°$; artificially rough bed. [Armanini and Scotton, 1992]

problem of steady front of real fluid propagating with uniform celerity .

The front of real fluid propagating with uniform celerity a, has been treated by Takahashi[1980]. Considering a reference frame ($X_r = x - at$), moving as the travelling celerity of the front a, the relative velocity is:

$$U_r = U - a = 0 \tag{7}$$

with respect to the moving frame, moreover, the profile is steady:

$$\frac{\partial h}{\partial t} = \frac{\partial U}{\partial t} = 0 \tag{8}$$

On substituting eqs.7) and 8) into eq.1), one gets:

$$g\cos\theta\frac{\partial h}{\partial X_r} = g\sin\theta - \frac{\tau_0}{\bar{\rho}\,h} \tag{9}$$

Takahashi obtained the solution of eq.9), assuming for τ_0 the expression provided by Bagnold grain-inertia theory:

$$\frac{\tau_0}{\bar{\rho}} = kU^2 \tag{10}$$

where k is a friction coefficient, calculated according to grain collision theory of Bagnold, but kept independent on flow depth h.

Under these hypotheses, eq.9) can be rearranged in the following form:

$$\frac{\partial\left(\dfrac{h}{h_\infty}\right)}{\partial(\dfrac{X_r}{h_\infty}\tan\theta)} = 1 - \left(\frac{h}{h_\infty}\right)^{-1} \tag{11}$$

here h_∞ is the asymptotic flow depth far upstream the front.

The solution of eq.11) is the following:

$$\tan\theta\frac{X_r}{h_\infty} = \frac{h}{h_\infty}\ln(1 - \frac{h}{h_\infty}) \tag{12}$$

The equilibrium profile predicted by eq.12) has been found in good agreement with experimental data of snout propagation of a debris flow by Takahashi [1980].

It should be noted however, that according to Bagnold theory, the friction coefficient k is proportional to h^{-2}. If so, eq.11) is replaced by the following equation:

$$\frac{\partial\left(\dfrac{h}{h_\infty}\right)}{\partial(\dfrac{X_r}{h_\infty}\tan\theta)} = 1 - \left(\frac{h}{h_\infty}\right)^{-3} \tag{13}$$

the solution of which is the following:

$$\frac{X_r}{h_\infty}\tan\theta = -\sum_{i=1}^{\infty}\frac{1}{3\,i+1}\left(\frac{h}{h_\infty}\right)^{3\,i+1} \tag{14}$$

In case of a debris current over fixed bed eq.12) and eq.14) have been checked by Armanini and Scotton [1993]. Experimental results show a marked disagreement with theoretical prediction by both theories.

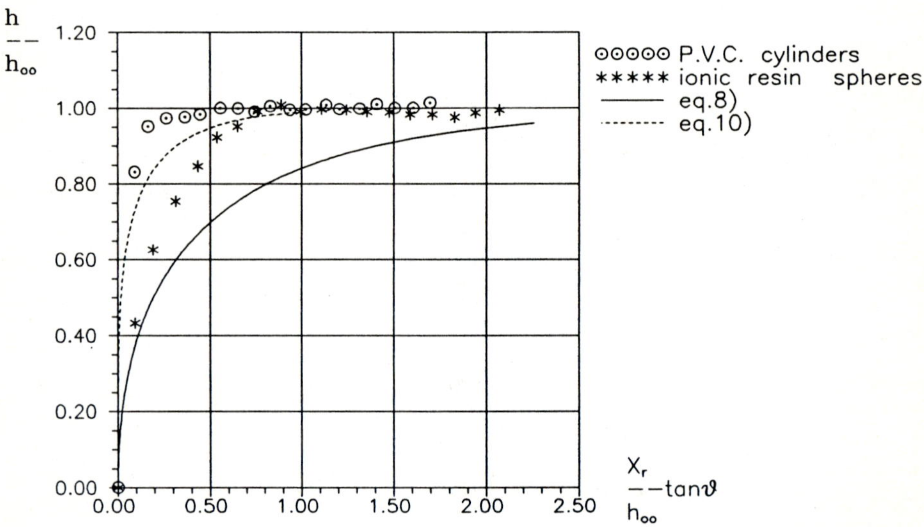

Fig. 4. Longitudinal profile of a debris current over fixed bed. Theoretical solutions [eq.12) and eq.14)] are compared with experimental data. $\theta = 11°$; $d = 2\,mm$; $\rho_s = 1300 kg/m^3$. [Armanini and Scotton, 1983]

From Fig 4 the different behaviour of experiments relevant to different materials suggests that the solution must depend also by other dimensionless parameters, like e.g. (d/h_∞), (λ/λ_∞), $(\bar{\rho}/\bar{\rho}_\infty)$ etc., where d is the grain size,

λ is the linear concentration of Bagnold, and the index ($_\infty$) is related to the far upstream uniform flow condition.

3. Two Kinds of Impact: Jet-Like and Reflected Wave

The experimental analysis [Armanini and Scotton,1992], suggests that basically two types of impact can be formed. In the first type, at the time of the impact the flow is completely deviated so as to form a vertical jet- like bulge. According to a second type, at the time of the impact a reflected bore is formed, which is going to propagate upstream.

The difference between the two impact features seems depend on the profile of the front at the time of impact respect to the equilibrium snout defined as above (eq.12 or eq.14).

Depending on the sign of right hand side of eq.1), the unsteady profile, relative to a coordinate system moving as the travelling celerity of the front a, tends to increase or to decrease in time. If friction forces prevail over gravity forces, the flow tends to decelerate and the profile tends to become deeper. In this case at the time of the impact a reflected wave (reflected bore) is forming. A more convincing explanation of this phenomenon can be given in terms of characteristics in a space-time plane, [Henderson, 1966]. The concavity of the characteristic relative to edge point of the profile is positive: at the time of impact the characteristics tend to envelop (fig. 5).

On the contrary, if gravity forces prevail over friction forces, the flow tends to accelerate and the profile tends to rise in time. In this case, at the time of the impact, a vertical jet will form. The characteristic relative to edge point of the profile presents a negative concavity: at the moment of impact the characteristics in a space-time plane diverge without enveloping (Fig.6).

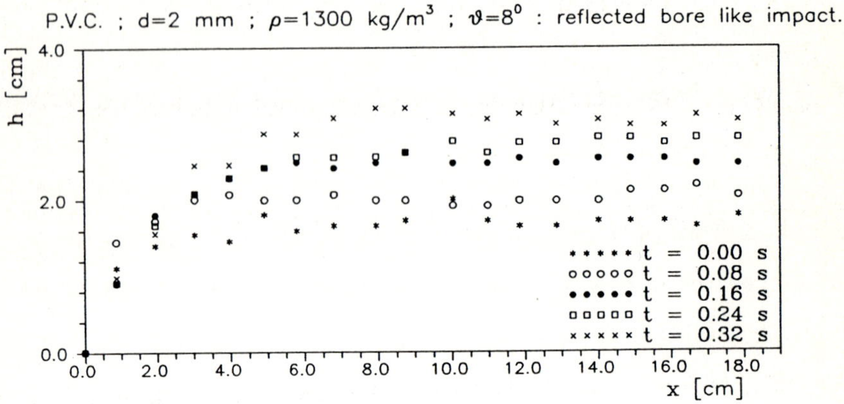

Fig. 5. Front profiles at different time, in an experiment where at the time of the impact a reflected wave is formed [Armanini and Scotton 1993].

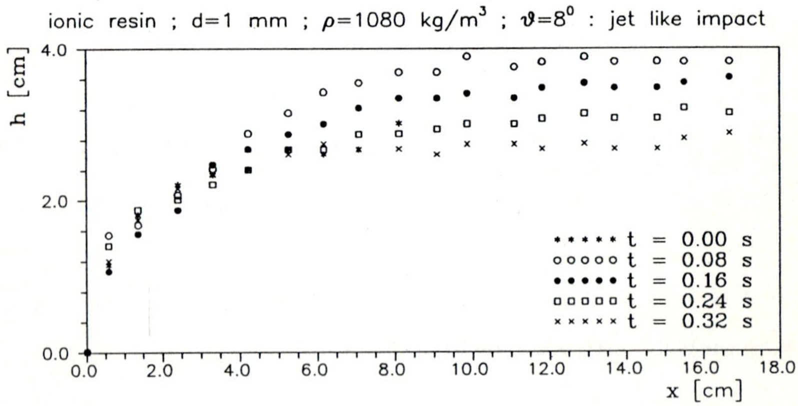

Fig. 6. Front profiles at different time, in an experiment where at the time of the impact a vertical jet-like bulge is formed [Armanini and Scotton 1993]

In Fig. 7 the velocity of increase of snout profile is plotted versus the bed slope for different experimental conditions. We note that:

- Zero increasing velocity corresponds to steady profile: the impact shape is uncertain;
- positive velocity corresponds to accelerating profile: in this case the impact is always of reflected wave type;
- negative velocity (decreasing profile in time) gives place to a vertical jet-like profile.

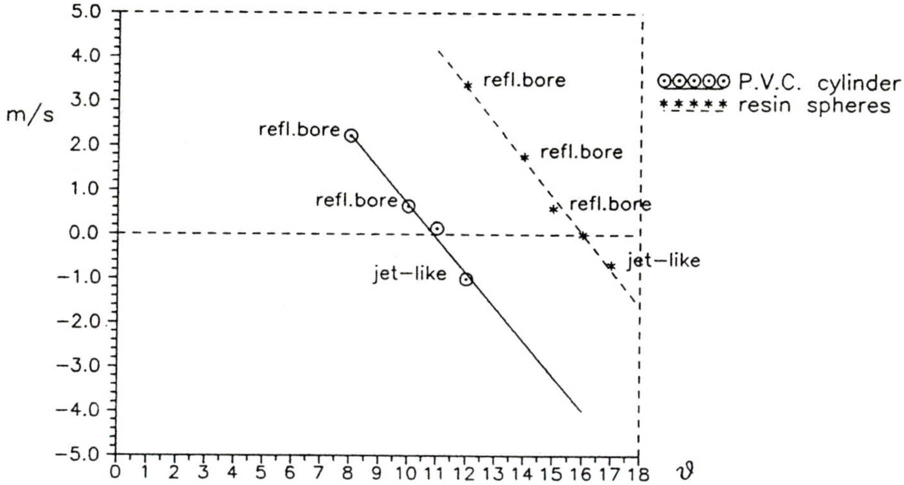

Fig. 7. Velocity of increase of snout profile, at the time of impact, versus the bed slope [Armanini and Scotton 1993].

4. Dynamic Impact Calculation

Referring to the sketch of Fig. 8, the dynamic impact can be calculated by applying the global momentum balance to the control volume $\forall_c$ with reference to a fixed coordinate system:

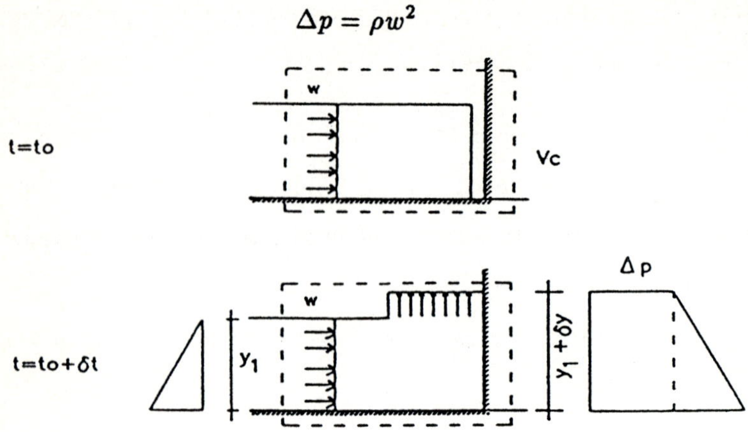

Fig. 8. Scheme of the dynamic impact and notations

$$\frac{1}{2}\gamma h^2 + \bar{\rho} h w^2 = \frac{1}{2}\gamma (h + \Delta h)^2 + \Delta p h + \frac{\partial}{\partial t}\int_{\forall_c} \bar{\rho} w d\forall \tag{15}$$

According to the two schemes of impact described above, the momentum balance on the wall can be rearranged in different way, leading to different results.

Under the <u>first hypothesis</u>, when the flow, at the impact time, is deviated along the vertical direction (Fig. 9), one can neglect the variation in time of momentum within the control volume.

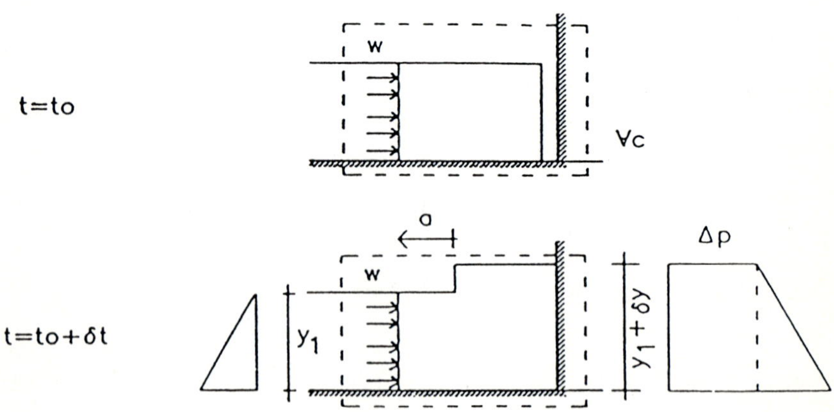

Fig. 9. Scheme of the dynamic impact, when at the time of the impact the flow is completely deviated so as to form a vertical jet-like bulge

The last term at right-hand side of eq.15) can be omitted, to obtain:

$$\Delta p = \rho w^2 \tag{16}$$

Under the second hypothesis, at the time of the impact, the fluid does not deviate in the vertical direction, but is reflected in the opposite flow direction, forming a reflected wave which propagates upstream.

In this case it is no more acceptable to neglect the time variation in the momentum balance (eq.15).

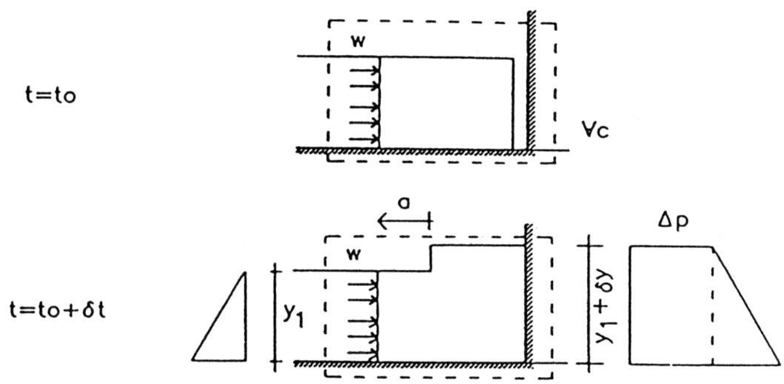

Fig. 10. Scheme of the impact model, when at the time of the impact a reflected surge is formed

Momentum balance in this situation (Fig. 10) yields:

$$\begin{aligned}\frac{1}{2}\gamma h^2 - \frac{1}{2}\gamma (h + \Delta y)^2 - \Delta p(h + \Delta h) &= -\rho h w^2 + \frac{1}{\delta t}\left(\rho w h(-\delta x)\right) \\ &= -\rho w h(w + a) \end{aligned} \tag{17}$$

or:

$$\frac{\Delta p}{\gamma} \simeq -\Delta h + \frac{w(w+a)}{g} \simeq \frac{w(w+a)}{g} \tag{18}$$

According to eq.16) and to eq.18) the dynamic impact forces are proportional to the square of front celerity w.

In Fig 11 some experimental results relative to this relationship are reported. From the figure one derives that, when light granular material is present (anionic resin 100%, $\rho_s = 1080\ kg/m^3$), the maximum pressure at the same front velocity is significantly reduced: $\Delta p/\gamma = 0.45 w^2/g$.

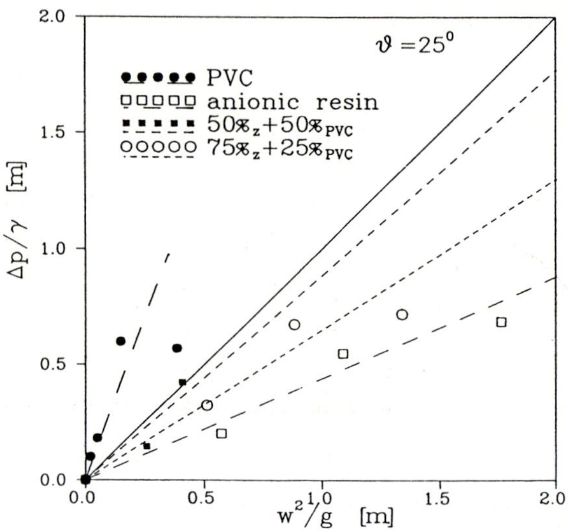

Fig. 11. Maximum pressure values ($\Delta p/\gamma$) versus the front celerity w, in case of a mixture of granular material of different composition; $\theta = 25°$; artificially rough bed. [Armanini and Scotton 1992]

On the contrary the dependence of over-pressure on front celerity is increasing with increasing concentration of cylindrically shaped PVC material. In case of PVC and water mixture, the maximum pressure per unit velocity is even larger than theoretical one ($\Delta p/\gamma = 2.2 w^2/g$).

In the first case in fact gravity forces prevails over friction forces and the snout is still accelerating at the moment of impact (see Fig. 6): a vertical jet-like bulge is formed. In the second case the internal friction is relatively higher, at impact time the snout profile is decelerating (see Fig.5) and the fluid does not deviate in the vertical direction, but is reflected in the opposite flow direction, forming a reflected bore which is going to propagate in upstream direction. Eq.18) is more proper for this kind of impact.

This difference between the two situations is likely due to the value of the internal friction, which, in case of cylindrical material, is much higher than that of clear water or of the material as discussed above.

On combining eq.6) and eq.16), one obtains the maximum theoretical pressure as a function of undisturbed flow depth, for the dam break of ideal fluid.

$$\frac{\Delta p}{\rho} = 4c_o^2 = 4gh_o \cos\theta \tag{19}$$

Equation 19) suggests that the maximum theoretical dynamic pressure, in case of non viscous fluid and horizontal bottom, is four times the undisturbed flow depth in the upstream reservoir.

Generalizing this result, one might infer that if the slope is such that the weight component in the flow direction is balanced by the resistance due to the roughness on the bottom, the maximum impact pressure is equal to four times the upstream depth.

One can perceive by intuition that the steeper is the slope, the higher is the front velocity and the higher the impact pressure; on the contrary, the milder is the slope (or the greater is the roughness of the channel) the lower is the front velocity; in this case the ratio between dynamic pressure and γh_0 will be less than 4.

Fig. 12 shows the results of experiments relevant to clear water. The maximum dynamic pressure is given as a function of the upstream water depth, at different values of channel slopes, in case of smooth bed. From the figure it indeed appears that the relation between the maximum pressure and the upstream hydraulic head is linear:

5. Conclusions

The results obtained suggest that the dynamic impact of debris flows on fixed structures does not depend on the height on the dam but rather on the front velocity. This may explain the dispersion in the practical formulas proposed by authors which have tried to correlate the dynamic pressure to the hydrostatic pressure.

It must be stressed, however, that often debris flow is originated by a natural dam break where the flow head corresponds to torrent depth [Hungr,O. et alii, 1984]. In this case also because the height of check dams is not much different from the torrent depth. If one calculates the dynamic impact force S on the basis of eq.19) in terms of hydrostatic pressure, one obtains:

$$\begin{aligned} S &= \Delta p \; h_0 + \tfrac{1}{2}\gamma h_0^2 \\ &= 4\gamma h_0^2 + 0.5\gamma h_0^2 \quad = 4.5\gamma h_0^2 = 9\,\frac{1}{2}\gamma h_0^2 \end{aligned} \tag{21}$$

The coefficient 9 in eq.21 is surprisingly close to the values recommended by some European torrent control authorities for design the structures against debris flow ($7 \div 11$ times the hydraulic pressure [Lichtenhahn, 1973]).

In real cases it is likely that instabilities and front curvatures might affect the validity of the one-dimensional theory. Hence the phenomenon of boulder expulsion does likely influence both the front celerity and the dynamic impact. Finally it is worth noting that accelerative phenomena are poorly reproduced in reduced scale models.

6. Acknowledgments

This research was supported by a grant from the C.N.R. G.N.D.U.C.I.. The writers are also indebted with ing. Marina Giusti, which has contributed to this research during the prepare on her degree thesis.

REFERENCES

Armanini, A., and Scotton, P., 1992, Experimental analysis on dynamic impact of a debris flow on structure. *6th INTERPRAEVENT 1992*, Bern 1992, Vol. 6.

Armanini, A., and Scotton, P.,1993, Experimental investigation on dynamic impact of a debris flow. *Quaderni del Dipartimento di Ingegneria Civile ed Ambientale dell'Università di Trento*, in press.

Aulitzky, H., 1990, Vorläufige Studienblätter zu der Vorlesung Widbach-u.Lawinenverbauug, Sommersem.1990. *Einverlag des Inst. für Wildbach u.Lawinenverbau, Universität für Bodenkultur*, Wien, Bl, Nr.2.6/12a 2.6/31.

Henderson, F.M., 1966. Open Channel Flow, *Macmillan Series in Civil Engineering.*

Hungr, O., Morgan, G.C., and Kellerhals,R., 1984, Quantitative analysis of debris torrent hazard for design of remedial measures. *Can.Geotech.Journ, Vol.21, pp.663-667.*

Lichtenhahn, C., 1973, Die Berechnung von Sperren in Beton und Eisenbeton, *Kolloquium on Torrent Dams ODC 384.3*, Mitteilungen der Forstlichen Bundes-Versuchsanstalt, Wien, 102. Heft, pp.91-127.

Scotton, P., and Armanini, A.,1992, Experimental investigation of roughness effects of debris flow channels, *6th Workshop on Two-phase Flow Prediction* ,Erlangen March 30-April.

Takahashi, T., 1980, Debris flow on prismatic open channel. *J.Hydraul.Div., ASCE 106*: 381-96.

DISCUSSION

Takahashi: 1. I think in the real debris flow, as soon as it strikes the dam wall, deposition takes place and it affects the impact force. I would like to hear your comments on the issue.

2. The impact force of the big boulder may be another very important factor to be taken into account. Which factor do you consider more important between the hydro-dynamic pressure and the impact of the boulder?

Armanini: As far as the first question is concerned, the possible deposition in front of the dam at the moment of impact is included in the scheme of *reflected wave* kind of impact described in Paragraph 3 of the paper.

As for the impact force due to big boulders I agree that in some cases the impact due to big boulders could increase the dynamic impact. I belive, however, that since the velocity of big boulders is generally less than the bulk velocity, this effect is of minor importance respect to the hydrodynamic pressure.

Printing: Weihert-Druck GmbH, Darmstadt
Binding: Theo Gansert Buchbinderei GmbH, Weinheim

Lecture Notes in Earth Sciences

Vol. 1: Sedimentary and Evolutionary Cycles. Edited by U. Bayer and A. Seilacher. VI, 465 pages. 1985. (out of print).

Vol. 2: U. Bayer, Pattern Recognition Problems in Geology and Paleontology. VII, 229 pages. 1985. (out of print).

Vol. 3: Th. Aigner, Storm Depositional Systems. VIII, 174 pages. 1985.

Vol. 4: Aspects of Fluvial Sedimentation in the Lower Triassic Buntsandstein of Europe. Edited by D. Mader. VIII, 626 pages. 1985. (out of print).

Vol. 5: Paleogeothermics. Edited by G. Buntebarth and L. Stegena. II, 234 pages. 1986.

Vol. 6: W. Ricken, Diagenetic Bedding. X, 210 pages. 1986.

Vol. 7: Mathematical and Numerical Techniques in Physical Geodesy. Edited by H. Sünkel. IX, 548 pages. 1986.

Vol. 8: Global Bio-Events. Edited by O. H. Walliser. IX, 442 pages. 1986.

Vol. 9: G. Gerdes, W. E. Krumbein, Biolaminated Deposits. IX, 183 pages. 1987.

Vol. 10: T.M. Peryt (Ed.), The Zechstein Facies in Europe. V, 272 pages. 1987.

Vol. 11: L. Landner (Ed.), Contamination of the Environment. Proceedings, 1986. VII, 190 pages.1987.

Vol. 12: S. Turner (Ed.), Applied Geodesy. VIII, 393 pages. 1987.

Vol. 13: T. M. Peryt (Ed.), Evaporite Basins. V, 188 pages. 1987.

Vol. 14: N. Cristescu, H. I. Ene (Eds.), Rock and Soil Rheology. VIII, 289 pages. 1988.

Vol. 15: V. H. Jacobshagen (Ed.), The Atlas System of Morocco. VI, 499 pages. 1988.

Vol. 16: H. Wanner, U. Siegenthaler (Eds.), Long and Short Term Variability of Climate. VII, 175 pages. 1988.

Vol. 17: H. Bahlburg, Ch. Breitkreuz, P. Giese (Eds.), The Southern Central Andes. VIII, 261 pages. 1988.

Vol. 18: N.M.S. Rock, Numerical Geology. XI, 427 pages. 1988.

Vol. 19: E. Groten, R. Strauß (Eds.), GPS-Techniques Applied to Geodesy and Surveying. XVII, 532 pages. 1988.

Vol. 20: P. Baccini (Ed.), The Landfill. IX, 439 pages. 1989.

Vol. 21: U. Förstner, Contaminated Sediments. V, 157 pages. 1989.

Vol. 22: I. I. Mueller, S. Zerbini (Eds.), The Interdisciplinary Role of Space Geodesy. XV, 300 pages. 1989.

Vol. 23: K. B. Föllmi, Evolution of the Mid-Cretaceous Triad. VII, 153 pages. 1989.

Vol. 24: B. Knipping, Basalt Intrusions in Evaporites. VI, 132 pages. 1989.

Vol. 25: F. Sansò, R. Rummel (Eds.), Theory of Satellite Geodesy and Gravity Field Theory. XII, 491 pages. 1989.

Vol. 26: R. D. Stoll, Sediment Acoustics. V, 155 pages. 1989.

Vol. 27: G.-P. Merkler, H. Militzer, H. Hötzl, H. Armbruster, J. Brauns (Eds.), Detection of Subsurface Flow Phenomena. IX, 514 pages. 1989.

Vol. 28: V. Mosbrugger, The Tree Habit in Land Plants. V, 161 pages. 1990.

Vol. 29: F. K. Brunner, C. Rizos (Eds.), Developments in Four-Dimensional Geodesy. X, 264 pages. 1990.

Vol. 30: E. G. Kauffman, O.H. Walliser (Eds.), Extinction Events in Earth History. VI, 432 pages. 1990.

Vol. 31: K.-R. Koch, Bayesian Inference with Geodetic Applications. IX, 198 pages. 1990.

Vol. 32: B. Lehmann, Metallogeny of Tin. VIII, 211 pages. 1990.

Vol. 33: B. Allard, H. Borén, A. Grimvall (Eds.), Humic Substances in the Aquatic and Terrestrial Environment. VIII, 514 pages. 1991.

Vol. 34: R. Stein, Accumulation of Organic Carbon in Marine Sediments. XIII, 217 pages. 1991.

Vol. 35: L. Håkanson, Ecometric and Dynamic Modelling. VI, 158 pages. 1991.

Vol. 36: D. Shangguan, Cellular Growth of Crystals. XV, 209 pages. 1991.

Vol. 37: A. Armanini, G. Di Silvio (Eds.), Fluvial Hydraulics of Mountain Regions. X, 468 pages. 1991.

Vol. 38: W. Smykatz-Kloss, S. St. J. Warne, Thermal Analysis in the Geosciences. XII, 379 pages. 1991.

Vol. 39: S.-E. Hjelt, Pragmatic Inversion of Geophysical Data. IX, 262 pages. 1992.

Vol. 40: S. W. Petters, Regional Geology of Africa. XXIII, 722 pages. 1991.

Vol. 41: R. Pflug, J. W. Harbaugh (Eds.), Computer Graphics in Geology. XVII, 298 pages. 1992.

Vol. 42: A. Cendrero, G. Lüttig, F. Chr. Wolff (Eds.), Planning the Use of the Earth's Surface. IX, 556 pages. 1992.

Vol. 43: N. Clauer, S. Chaudhuri (Eds.), Isotopic Signatures and Sedimentary Records. VIII, 529 pages. 1992.

Vol. 44: D. A. Edwards, Turbidity Currents: Dynamics, Deposits and Reversals. XIII, 175 pages. 1993.

Vol. 45: A. G. Herrmann, B. Knipping, Waste Disposal and Evaporites. XII, 193 pages. 1993.

Vol. 46: G. Galli, Temporal and Spatial Patterns in Carbonate Platforms. IX, 325 pages. 1993.

Vol. 47: R. L. Littke, Deposition, Diagenesis and Weathering of Organic Matter-Rich Sediments. IX, 216 pages. 1993.

Vol. 48: B. R. Roberts, Water Management in Desert Environments. XVII, 337 pages. 1993.

Vol. 49: J. F. W. Negendank, B. Zolitschka (Eds.), Paleolimnology of European Maar Lakes. IX, 513 pages. 1993.

Vol. 50: R. Rummel, F. Sansò (Eds.), Satellite Altimetry in Geodesy and Oceanography. XII, 479 pages. 1993.

Vol. 51: W. Ricken, Sedimentation as a Three-Component System. XII, 211 pages. 1993.

Vol. 52: P. Ergenzinger, K.-H. Schmidt (Eds.), Dynamics and Geomorphology of Mountain Rivers. VIII, 326 pages. 1994.

Vol. 53: F. Scherbaum, Basic Concepts in Digital Signal Processing for Seismologists. X, 158 pages. 1994.

Vol. 54: J. J. P. Zijlstra, The Sedimentology of Chalk. IX, 194 pages. 1995.

Vol. 55: J. A. Scales, Theory of Seismic Imaging. XV, 291 pages. 1995.

Vol. 56: D. Müller, D. I. Groves, Potassic Igneous Rocks and Associated Gold-Copper Mineralization. 2nd updated and enlarged Edition. XIII, 238 pages. 1997.

Vol. 57: E. Lallier-Vergès, N.-P. Tribovillard, P. Bertrand (Eds.), Organic Matter Accumulation. VIII, 187 pages. 1995.

Vol. 58: G. Sarwar, G. M. Friedman, Post-Devonian Sediment Cover over New York State. VIII, 113 pages. 1995.

Vol. 59: A. C. Kibblewhite, C. Y. Wu, Wave Interactions As a Seismo-acoustic Source. XIX, 313 pages. 1996.

Vol. 60: A. Kleusberg, P. J. G. Teunissen (Eds.), GPS for Geodesy. VII, 407 pages. 1996.

Vol. 61: M. Breunig, Integration of Spatial Information for Geo-Information Systems. XI, 171 pages. 1996.

Vol. 62: H. V. Lyatsky, Continental-Crust Structures on the Continental Margin of Western North America. XIX, 352 pages. 1996.

Vol. 63: B. H. Jacobsen, K. Mosegaard, P. Sibani (Eds.), Inverse Methods. XVI, 341 pages, 1996.

Vol. 64: A. Armanini, M. Michiue (Eds.), Recent Developments on Debris Flows. X, 226 pages. 1997.